Inner Powers

Maximizing Performance Through Inner Power Development

In this book, Master Del Pe draws on his unparalleled experiences and training with four advanced Sages from different spiritual lineages, augmenting his technical education as a mechanical engineer, a master in martial arts and an expert in esoteric science. Using his engineer's and business mind to research, validate and apply abstract principles from the teachings of the Ageless Wisdom and esoteric sciences, which have been held secret for many millennia, he has formulated a newer presentation of these timeless truths into simple, easy-to-learn techniques on how to:

- Improve vitality and self-manage your health
- Culture your emotions and mind for maximum power & performance
- Develop your psychic powers for better problem-solving & decision-making
- Balance your material life with spiritual development
- Bring your success into fulfillment

This first volume of three offers a higher holistic way of viewing ourselves and our true human nature to be able to serve our families, groups, nations and the world better. This book prepares one to study the advanced Volumes II & III of *Inner Powers*.

Inner Powers

Volume 1
Maximizing Performance Through Inner Power Development

Master Del Pe

MDP
GLOBAL RESOURCES

Published in the United States by
MDP Global Resources®
P.O. Box 7947
The Woodlands, Texas USA 77387
Tel: (800) 352-6014 • (936) 273-9153 • Fax: (936) 273-9230
mdp@mdpglobal.com • www.mdpglobal.com

ISBN 0-9717676-0-2

Production Team
Jenna Wayne: Editor, Text Layout and Clairvoyant Consultant
Rafael Perez: Graphic Artist and Clairvoyant
Liliana Chethuan: Clairvoyant Consultant
Carolyn Wilder: Art Editor, Text Review and Logistics
Kym Maier: Photographer
Rafael Perez & Silver Rock Productions: Cover Design
Susan Powell, Suzanne Jarvis, Kain Sanderson: Insight and Review
Bibiana Vasquez, Lorena Dini and MDP SOIP Group: Technical Support

Dedication

To Humanity on its path to spiritual expansion
To the readers who are working for greater performance
To those who seek a better quality of life

And

To my family, students and co-workers
because you have co-created this book with me

And

To MCKS, MMCN, SHJ and BbT

Because I am so grateful for and blessed with
your mentorship and brotherhood

INNER POWERS
Volume I
Manual for Maximizing Performance Through Inner Power Development

Table of Contents

Introduction.. page i

Part I: What is power & its sources?

Section A: What is Power?............................ page 1

General concept and sources of power.............. page 2
The esoteric nature of power.......................... page 6

Section B: Inner Powers................................ page 10

Sources and derivatives of inner powers............ page 10
Stages of human development......................... page 14

Section C: The Inner Anatomy of a Human Being............................... page 16

The different human energy bodies.................. page 17
Components of the human energy system........... page 18
Human energy centers or chakras.................... page 28

Part II: Inner Powers & Their Types

Section A: Power of vitality: Principles & Techniques.................................. page 43

Sources of the power of vitality...................... page 44
Special exercises and breathing techniques.......... page 51

Section B: Emotional Power & Intelligence: Principles & Techniques.................. page 89

Ramifications of emotional intelligence............. page 90
Techniques to build emotional power & intelligence.. page 97

Section C: **Mental Power: Principles & Techniques**.................................. page 115

Procedures for programming and materializing goals.......................... page 125
Ramifications of mental power............ page 130
How the mind works and practices to develop it .... page 138

Section D: **Psychic powers: Principles & techniques**.................................. page 179

Psychic power experiments................ page 182
Types of psychic powers..................... page 194
Basic requirements for psychic power development.......................... page 195
Chakras for psychic power development.................................. page 205
Practices for psychic power development.................................. page 208

Section E: **Spiritual Powers: Principles & Techniques**.................................. page 287

Questions about spiritual powers.......... page 290
Major sources of spiritual powers and how to connect with them............. page 309
Power of virtues and culturing the personality............................... page 338

Part III: **The Importance and Relationship of Inner Powers to Evolution, Quality of Life and Practical Performance**.................................... page 407

Maximizing performance equations.................. page 410
Strategies for maximizing performance.............. page 412
Recommended weekly programs..................... page 424
Schedules for maximizing performance program... page 429

Glossary.. page 434

Index of Sources of Selected Figures.................................. page 440

Table of Diagrams & Illustrations

Figure 1: Trinities in Different Spiritual Traditions............ page 7

Figure 2: The Human Energy System .... page 17

Figure 3: Anchor Points of the Soul in the Different Bodies.. page 25

Figure 4: Schematic Diagram of the Spirit-Soul-Personality Relationship............................... page 27

Figure 5: Sample Chakras – 3 Layers............................ page 28

Figure 6: Locations of Crown, Ajna, Heart and Navel Chakras...,page 30

Figure 7: Cross-Section of Etheric Navel Chakra.............. page 31

Figure 8: Sample Chakra in Motion.............................. page 32

Figure 9: Crown and Heart Symbols............................... page 36

Figure 10: Heart and Alta Major Symbols........................ page 38

Figure 11: Air Constitution and Earth Energy.................... page 47

Figure 12: MDP Body Therapy Exercises........................ page 58

Figure 13: Inner Power Development Procedure................. page 70

Figure 14: *Anjali Mudra*.. page 71

Figure 15: Proper Abdominal Breathing........................... page 76

Figure 16: Positions for Meditation and Breathing Practices....................................... page 77

Figure 17: Combination Breathing................................. page 80

Figure 18: MDP Synchronized Breathing Technique............ page 84

Figure 19: Amplified Portrayal of the Substance of the Virtue of Fearlessness and Fear Entity as Seen by a Clairvoyant.................................. page 100

Figure 20: Amplified Portrayal of the Substance of the Virtue of Love and the Substance of Lower Passion.. page 101

Figure 21: Amplified Portrayal of Substances of Prosperity.. page 102

Figure 22: Breathing Technique to Develop Love-Intellect Faculty.. page 174

Figure 23: Schematic of the Chakra Cross-Section Showing the Energy Screens/Sieves.................page 200

Figure 24: Different Locations of the Chakras..................page 207

Figure 25: Internal Martial Arts Hand Sensitizing Technique...page 213

Figure 26: Auras of Medications...................................page 219

Figure 27: Seeing the Aura of Your Fingers.....................page 224

Figure 28: The Different Conditions of the Chakra's Energy Screens/Sieves..................... page 231

Figure 29: Energy Telekinesis.................................... page 270

Figure 30: Starting to Go Out of Body...........................page 279

Figure 31: Schematic Diagram of the Human Triplicity.........page 304

Figure 32: Earth Being as a Macrocosm.........................page 314

Figure 33: Cobra as a Symbol of the *Kundalini*.................page 321

Figure 34: Schematic Illustration of the *Kundalini* and the Medical Caduceus............................page 321

Figure 35: Hand *Mudras* for Meditation and Circulation of Energy.................................page 333

Figure 36: Integrated Life Planning.............................. page 421

Introduction

They say that those who don't learn from history are doomed to repeat it. But what if, for all of humanity's efforts to understand history in order to understand itself, we've been looking through an incomplete prism or wrong lens for the lessons?

World history revolves around power – power of men, women, clans, cartels, corporations, nations, religions, conquerors and leaders. The triumphs and errors of history all feed on the inevitable lure and uses or misuses of power. Even our daily personal lives – whether we realize it or not – are shaped by the possession or lack of power.

But have you ever stopped to wonder where power really comes from, why some people have it and others don't, and whether events of world history are perhaps shaped by a power most people are unaware of? Is it possible that the workings of the outer, visible world are actually shaped by inner powers, which operate with a different perspective, a different sense of time and a different type of reality? Is it possible that the powers of human leaders are just fleeting, impermanent reflections of deeper inner powers latent in every person, which are mirrored and reflected from a great unshakable abstract source of power and occasionally tapped by those whom civilization considers brilliant?

What characterizes this brilliance? Brilliance is characterized by the times. Today we consider Bill Gates, in his own world of technology and leadership, and Jack Welsh, in his business leadership, extraordinary. Early in the 20th century, Albert Einstein, Madame Curie and Thomas Edison were considered brilliant. And down through the ages, Shakespeare, Beethoven, Mozart, Francis Bacon, Galileo, the Lord Jesus the Christ, Confucius, Socrates, Lao Tzu and the Buddha have shined as beacons of brilliance and enlightenment. The impulses and sensibilities of their respective ages in history shaped society's perception of them, but what also made them stand apart then and now – even thousands of years after they walked this planet – is their very timeless principles and creativity. Most of them were wise human beings in eras without universities, and their wisdom endures even in this modern age of rapidly unfolding and expanding knowledge.

So what makes them endure? What made them wise? What made them extremely creative? The answer is simple: inner powers. All of them shared innate inner faculties and secrets of vitality, emotional intelligence and magnetism, highly developed mental powers, psychic powers and spiritual powers. They were all able to tap into an abstract, intuitional plane – which yet awaits the majority of humanity's discovery – to bring high ideas down into the concrete realms for the good of civilization's development.

There are inner reasons why these particular individuals discovered and possessed inner powers, but some of their skills and faculties are not out of reach for the rest of us. Great business leaders and scientists can create an idea – and with one-pointed focus and a rhythm of building – make it a physicalized reality. Mothers sense when their children encounter danger through natural instincts. Our instincts react without thinking if we step off a curb into the path of an oncoming speeding car. These are all everyday expressions of the magnificent inner powers latent in every human being.

So, the good news is that everyone is capable of unleashing their inner powers through a systematic learning process, and it's never too late to begin!

By mastering the timeless tools and techniques for inner power development available in this book, you can enjoy levels of performance, success and fulfillment you may never have imagined possible. These teachings from what is called the "Ageless Wisdom" – the wisdom of the Sages – will help you accomplish in a few years what other people cannot achieve in a lifetime. When you apply the techniques of inner powers to your life with discipline, you will one day find yourself in the portal of higher evolution and enlightenment, able to offer greater service to the world.

The knowledge you will gain from this book will give you the inner power to improve your personal and professional life tremendously through:

1. Mastery of methods that increase your vitality fast and rejuvenate your health for longevity and productivity
2. Development of emotional intelligence and mental powers for better performance in leadership and personal career
3. Proficiency in the proper use of psychic powers for visionary leadership, intuitive management and greater effectiveness in decision-making and problem-solving
4. A scientific and practical approach to spiritual unfoldment
5. A deeper insight into your true inner nature and spiritual anatomy as a human being, revealing a new psychology of life
6. A better understanding of your past, present and future and how to harmonize this knowledge in order to transmute your experience into wisdom
7. Esoteric Science information to demystify life's experiences and illuminate some gaps of consciousness not explainable by religion, education and science
8. Concrete procedures and strategies to deliberately attain the various levels of spiritual illumination
9. A systematic approach to integrate material and spiritual life, which can bring you from success to fulfillment

The teachings in this book have either been discovered by me or came through my extensive training with my earlier spiritual mentors. If you put them to regular use, your life can be positively enhanced in all aspects – family and home, career and work, social and environmental contribution, spiritual and health – and you can live in balance with "one foot on Earth and one foot in Heaven."

Welcome to your new journey of unfoldment along the path to inner powers.

Master Del Pe

Before we begin, here are some icons to help you navigate through this book. Each icon notes a particular technique, recommended practice schedule, precautions when using techniques, experiments and pearls of wisdom.

This icon appears whenever a new technique is introduced.

This calendar icon appears whenever a schedule for practicing a technique is recommended.

This icon notes a precaution regarding the practice of a technique.

This icon notes an experiment you can try to test a principle proposed in this book.

This crown icon appears wherever a pearl of wisdom is offered for contemplation.

Part I
What is Power & its Sources?

Section A: What is power? .... page 2

The general concept & sources of power .... page 2
The esoteric nature of power .... page 6

Section B: Inner Powers .... page 10

Sources and derivations of inner powers .... page 10
Different stages of human development .... page 14

Section C: The Inner Anatomy of a Human Being .... page 16

The different human energy bodies .... page 17
Physical body
Vitality or etheric body
Emotional or astral body
Mental body
The spiritual antenna
The Soul
The Spirit

The human energy centers or chakras .... page 28
Functions and descriptions of energy centers
Current forms and features of the
chakras of human beings
Approximate chakra sizes of human beings

Section A
What is power?

The general concept and sources of power

Throughout human history, the subject of power has not been scrutinized or researched as deeply as love or intelligence. It's hard to find any books on the science of power. Nonetheless, power is most sought after and enjoyed by great leaders in many fields, whether they are aware of it or not. However, many people and groups misunderstand the concept of handling power properly. Several stories and tales indicate the rise and fall of dynasties, empires, organizations, families or individuals because of misuse of power. Also, the decline of many leaders, organizations and nations is due to the fading of their powers. History has taken note of the decline of the ruling power of Egypt, Rome, Spain, England, Japan and Russia. Now the United States is following in the footsteps of these great empires, being challenged in its position as a world power.

In a more personal context, power can be expressed on many levels as stamina, strength, vigor and material, emotional, mental and spiritual influence and authority or supremacy. Many people associate these types of power with political, financial, religious and educational status, but these are just externalizations of inner faculties and expressions of aptitudes of individuals or groups of people. Let's look at some historical examples of inner powers in order to understand them further. Sir Isaac Newton had concrete mental power to express the natural laws as concrete equations whereas Lao Tzu had the abstract mental power to relate nature to daily life situations. The Christ had the power of love to teach the world, and Buddha had the power of wisdom to explain the path of enlightenment.

Power can be expressed through individuals, groups or organizations, nations or groups of nations. A study of different cultures reveals several sources of power:

Sources of Power
Power of Lineage
Power of Office or Position
Power through Expertise & Education
Power through Social Intelligence
Media Power
Physical Power
Team or Group Power
Financial Power
Power of Nature
Inner Powers

Let's examine each source of power.

Power of lineage

This type of power comes through a line of succession brought by "inheritance". Queen Elizabeth II, the Pope, the Kennedy Family and members of political parties possess this type of power. This power asserts itself depending upon the social acceptance and degree of influence of the lineage.

Power of office or position

An individual or group holds this type of power as long as they fill an office in an organization or in society. The office of the President of the United States is one of the most powerful positions in the world today. When one becomes the US President, the person holds great power. But when that President's term has ended, the ex-President becomes a normal citizen again.

Power through expertise and education

If you become an expert or highly educated specialist in any particular area, that's power, and the rarer and more required your feat or product, the greater the power. Extraordinary entrepreneurs, businesspeople and leaders of companies like Microsoft, IBM and GE who are experts at the cutting edge of their trade or business fall into this category. Many world-renowned scientists like Albert Einstein, Edwin Hubble and Thomas Edison also fall into this category.

Power through social intelligence

Some people are greatly liked by others or are cherished by multitudes. Their ability to connect to other human beings emotionally and to empathize with their needs and feelings make them socially magnetic. This is one of the sources of power in leadership, which we call mass appeal or the quality of being likable. Princess Diana possessed this quality more than her husband, Prince Charles, who has power through his lineage. Socially cultured individuals also have training to apply this type of power. Eleanor Roosevelt is an example: She enjoyed social influence and utilized this power to develop social projects. Many great leaders in many fields of service employ the power of social intelligence to magnetize their supporters.

Media power

Multimedia promotions, advertising and exposure have been utilized by many leaders, corporations, organizations and nations to empower an individual, a product, a business service or an ideal. The power of the media has been exploited to give power to wrong goals, ends and personalities. This same power has also been used to promote celebrities to a high level of social influence, including sports stars. The film industry and media have used this power to affect the world with new consciousness and attitudes more than any other modern group or instrument. The rise and fall of many powerful people and groups depend on public opinions generated by the media. The psychological effect of the September 11, 2001 Twin Towers incident in New York City would not have been nearly as powerful – with the extent of global impact – without the power of the media.

Physical power

Sports personalities and Olympic athletes embody physical power. Strength, vitality and stamina are their main displays of power. Even though great athletes also utilize mental will-power, physical feats and power are still their primary domain. In human beings, physical power does not last for a long time because it declines with age and poor health. This is often the case of sports heroes like Mohammad Ali, who was the king of boxing, but whose physical power faded away when he contracted Parkinson's disease. Another aspect of physical power is the power of good looks, physical beauty and sex appeal. Many prominent people with other types of power like position, money or expertise have fallen enslaved by the lure of physical beauty and sexuality, surrendering their careers, wealth and reputation. Of all the powers, physical power disappears most easily.

Team or group power

This is a true source of power and is now being applied successfully in business. Teamwork brings tremendous power and achievement, more than the sum total of the members' individual strengths. In modern global business and markets, even big corporations need to be members of world business alliances to be able to survive and compete internationally. Business organizations are excellent embodiments of this group organizational power. Teamwork is a must to stay in power in this century.

Financial power

Financial power has great influence on almost every human being, group, organization and nation in our modern life. This power is brought or affected by almost all the above mentioned powers. Although it is a temporary source of power, it is a major factor triggering many great leaders to work for good or bad ends and groups.

Power of nature

The Earth operates according to natural laws, some discovered – like the Law of Gravity – and some still hidden and unnoticed by many. The forces of the Earth, like the weather, earthquakes, volcanic eruptions, storms, hurricanes, fires and floods, demonstrate this power. There are also invisible forces of creation that serve as destroying, constructing or preserving agencies of the visible and invisible aspects of life and forms. There are at least 7 types of qualifying forces called the 7 Rays of Creation by the esoteric sciences. These forces and energies and their roles and activities are discussed in detail in *Inner Powers*, Volume II, *Manual for Spiritual Technology*.

Inner powers

In the political arena, many events in history like the World War II conflict between the Allied and Axis powers are but an extension of eons of time balancing the inner powers of hidden forces. This conflict of Invisible Forces is also manifested in the many wars and struggles fought among Islam, Judaism and Christianity, expressed as crusades or holy wars. At the heart of politics, religion and business lies a seed of power for either good or bad depending on the intentions and motives behind the use of powers. Inner powers lie behind all visible external powers. We will discuss inner powers in detail throughout the rest of this book.

The esoteric nature of power

Presently, the most powerful nations and organizations are those whose leaders are advanced visionaries and thinkers. Wealthiest and handling world-class power, even their powers aren't permanent. They belong to the personality aspects of human beings and are transitory compared to inner or spiritual types of powers, which are more permanent because they derive from the Soul or Spirit, from greater invisible Beings or from the Divine Macrocosm. Advanced human beings know how to access and wield these inner powers. Human expressions of extraordinary power are the external ramifications of hidden forces and energies that come from a greater planetary source. These esoteric sources will be discussed separately in Volume II of *Inner Powers*.

Three inner aspects shape outer life – all life: 1) Will-power 2) Love and 3) Active Intelligence. Power is the first quality and aspect of the Macrocosm or Divine Deity expressed as Omnipotence or Will-Power, the second aspect being Love and the third, Active Intelligence. There is a spiritual belief that human beings are made in the likeness and image of God or the Creator. This doesn't literally mean the physical form, but the divine qualities, which are omnipotence (all-powerful), all-loving and omniscience (all-knowing). In some religions, these are Divine qualities of God personified. Christians believe in the Trinity: God-the-Father, God-the-Son and God-the-Holy-Spirit, three Divine qualities in one God. In Hinduism, there are also three divine expressions of the Divine Deity: *Shiva* is the Divine Destroyer, *Vishnu* is the Divine Preserver and *Brahma* is the Divine Creator.

Trinities in Different Spiritual Traditions

Figure 1

If you observe the progress brought by the evolutionary process, nature demonstrates the above principle of the Trinity: In order to create new forms, there is a need to destroy the old, obsolete forms using power. Then new forms and consciousness are constructed using creative intelligence, after which they are preserved through love. These three qualities and aspects of the Macrocosm or Deity are the sources and stimulants of humanity's powers and their ramifications. For example, the three major divine qualities can be broken down as follows:

- Will-power can be expressed as the powers of stamina, vitality and strength, endurance and persistence. In its other form, will-power can also be used for constructive destruction. There is an old saying that expresses this:

> You cannot build new temples without destroying the old ones first.

- Love can be expressed as the powers of compassion, understanding, unity, peace, altruism and benevolence.
- Active intelligence demonstrates through the powers of the mind, creativity, innovation and science.

The ability to become powerful and to be able to sustain it depends largely on how much of the three qualities of Will-Power, Love and Active Intelligence can be mastered singularly or in combination. The more integrated these qualities in a human being, the more power he or she has. Let's take the example of leadership:

- The best motivating and inspiring leaders are those who have the power of love or the heart. They are most loved and adored by their followers, like the Christ or the Buddha and other founders of religions.

- The leaders with will-power and political influence are rulers and have authority over nations.
- The leaders with creative intelligence have the power of money, resources, arts and technology.

There are also two major sources of influence in the development and utilization of power in humanity: the two invisible forces of evolution and involution. Both of these comprise major sources of power for humanity to use for either good, constructive goals or for what we call evil, perpetrated by those who are advocates for negative ends or involution. This is similar to, but not the same as, the terms religions apply. Human activities that advocate inclusiveness and expansion of consciousness and adhere to universal values are empowered by the power of light, the forces of evolution. The human goals and work that solicit separativeness, selfishness and that limit humanity's consciousness and free will attract the power from the involutionary forces, the "forces of darkness". The motives, intentions and quality of life one lives determine what side of power is being employed.

> Contrary to what one might expect, the true esoteric mastery of Inner Powers depends on the mastery of the virtues of benevolence, wisdom and universality.

The Christ, the Buddha, certain great prophets and saints are historical examples of these virtues.

Section B
Inner Powers

The single most important factor that can bring a person to great heights of power is inner power – be it that of a leader, a powerful official in government or a powerful head of a lineage. Take, for example, the life of Mahatma Gandhi, an Indian hero of small physical stature. His power came not just from his innate leadership skills, unwavering determination and mental stamina, but also from his inner power that accompanied serving a great cause and seeing the great vision at the end. This developed tremendous will-power in Gandhi to finish his spiritual duty – the same core of inner power that drove Martin Luther King to fight for human rights until the very end of his life through his servant and sacrificial leadership. The greatest human beings with the most number of followers have been the great founders and leaders of major religions – the Buddha, the Christ, the Mohammad, the Lord Krishna of India and the great prophets of the Jews like Moses. All of them had one thing in common: They all possessed inner powers. They did not acquire power through media exposure or formal education. It was their inner powers that constructed their path and the livingness of their work until today, still surviving thousands of years later.

Most of the well-remembered and influential beings in our world today have been associated with a religious kind of spirituality, but they are beyond specific religions; they no longer belong to the religions they founded. I see that their true religious expression of spirituality was their ability to unfold and express the highest truth about life, their ability to sacrifice their lives for a greater cause, for a bigger evolutionary purpose and their ability to bring long-term, permanent change for life on Earth – using inner powers, faculties and principles empowered by their Spirits or Divine Principles.

It is these inner powers that made the Christ, the Buddha and Great Sages different from today's most powerful leaders, politicians, scientists, executives and scholars. Among these Great Ones, money was not an issue and a diploma from a university was not available. In our modern times, these great spiritual leaders continue to powerfully influence the lives of the most important and wealthy people today even though they are physically absent. This means that even modern famous and influential personalities with money, reputation, media standing, political position, reputation or scientific expertise have not surpassed the magnitude of influence and power of these earlier leaders.

There are also many advanced Sages and many spiritual world servers working incognito on great global projects, whose intention it is to not be publicly known. In most cases, our great modern leaders like Abraham Lincoln, George Washington, Franklin Delano Roosevelt, Winston Churchill, General Bolivar of South America, Sir Isaac Newton, Albert Einstein, Madame Curie, Henry Ford, Lee Iacocca, Jack Welsh, Bill Gates and Oprah will probably be remembered for a few hundred years. We know that these contemporary great leaders tend to fade out as they retire or as their names don't get mentioned as much in the media. What is your hunch or prediction? How long will these great personalities be remembered? Let time be the yardstick.

Whether these famous people will be remembered over time remains a question, but some of them provide useful illustrations of external powers that mirror inner powers.

Individual Example	External Power	Equivalent Inner Power
Oprah Winfrey & Princess Diana	Power of charisma & ability to be liked by multitudes	Emotional power & magnetism
Bill Gates	Power of expertise in computer technology & financial power	Power of innovation & mental creativity, combined with altruism
Mother Teresa	Power of spiritual popularity & charismatic leadership	Servant leadership using the power of the heart & the power of benevolence
Sir Isaac Newton	Power of science	Excellent concrete mental intelligence
Albert Einstein	Power of science	Integration of abstract & concrete mental intelligence
Lao Tzu	Power of a Sage	Power of inner understanding & abstract mental intelligence
Theodore Roosevelt	Political power & leadership	Will-power, stamina & the virtue of discipline

The Christ	Power of a world teacher	Power of Universal Love & leadership through service & example
The Buddha	Power of enlightened leadership	Power of wisdom & righteousness

A teacher once said: "The power of true magnetism and the source of influence is not the ability to use gimmicks artificially or to use just physical looks to impress." It is the inner magnetism and presence brought in by the power of the Spirit or Soul down pouring into the personality that brings true inner power and attractiveness. Some people express this as a powerful presence or an inspirational feeling, even without normal communication channels. Some great Souls can inspire masses of people without even opening their mouths. In their silence or unwavering stillness, people enclosed in their auras are motivated and transformed. Christianity offers a good example of this: When Jesus was in a crowded room, a woman who had been sick and bleeding for many years wanted to reach him for healing. Because of the thickness of the crowd, she was only able to touch the hem of His robes – and she was healed, and Jesus exclaimed: "Behold, power has gone out of me!" On my trips in India, I also observed many devotees experiencing ecstasy and extreme joy as they sat in front of their spiritual guru, totally surrendered and devoted to their leader, seeing him not only as a personality but also as a great spiritual Sage.

The extent to which inner power is interpreted and sometimes misinterpreted is so extreme and diverse that I will elaborate on the different sources and derivatives of inner power in more detail in succeeding sections.

Inner Powers
First power: Vitality, resulting in physical stamina
Second power: Power of the emotions and emotional intelligence
Third power: Mental powers and intelligence
Fourth power: Psychic powers
Fifth power: Spiritual powers
Sixth power: Integration of any and all of the above powers

Different stages of human development

Humanity has evolved and is still evolving its inner powers through different stages.

First stage: Early humans

Early humans, due to their unevolved emotional and mental development, depended on their instinct to survive through physical instinct, stamina, vitality, strength and toughness. Their power lay only at the physical level.

Second stage: Average human beings

Average human beings, belonging to the masses of humanity, employ the power of emotions through lower passion, in addition to improved vitality and physical endurance.

Third stage: Intelligentsia

Intelligentsia, both mystical and mental types, have started to employ not only emotional intelligence and passion, but also the power of the concrete, logical mind. The basic psychic powers are also already being utilized by this group.

Fourth stage: Advanced human beings

Advanced human beings use their abstract and concrete mental faculties in addition to their emotional intelligence and power of love. They already have a sufficient development of psychic powers, whether they are aware of it or not. Spiritual powers are starting to be employed as a guiding tool in their advanced work and service. Most of them, especially the mental type, are starting to integrate the power of the mind, emotions, vitality and the physical body.

Fifth stage: Saints and Sages

Saints and Sages are very advanced human beings with highly developed inner powers. Saints and Sages, especially the mental type, utilize all the physical, vitality, emotional, mental, psychic and spiritual powers, though a large percentage of the mystical type of Saints and Sages still do not emphasize the physical aspects of their health due to older traditional concepts of abstract spirituality denouncing the material aspects of life. This is discussed in great detail in Volume II of *Inner Powers.*

In order to understand the sources of external and inner powers, and their relationship, we have to study and illustrate different esoteric aspects of a human being. There is an invisible human constitution, not recognized by formal education, unfamiliar to science and unknown to most human beings at this time, that serves as a mechanism where faculties, aptitudes, attitudes and virtues are fabricated. This human constitution is composed of many sheaths of invisible auras that process, contain and exude different types of consciousness and intelligences. Let's go to the next section in order to continue our journey to new discoveries.

Section C
The Inner Anatomy of a Human Being

Science and education have laid emphasis on the study of the physical aspect of the human being, with extensive research on atomic and sub-atomic particles, the physical anatomy and the physical cosmos. This is an impressive and remarkable achievement of humanity. In addition to the scientists and educators, other groups are also contributing on various levels:

1. Psychologists working on the emotional and mental aspects, but still focusing on the physical brain and nervous system
2. Energy healers mastering techniques addressing the vitality and energetic causes of diseases
3. Religion and their leaders placing importance on divine faith, doctrines of redemption and a righteous way of life
4. Spiritualists and mystics who are aware of the nature of the subjective realms and spirituality but who lack scientific tools and guidance to progress faster and with practicality

Unknown to most groups and the public are many spiritual teachers and esoteric scientists who are also conducting their own research and advanced studies on newer practical methods to bring spiritual unfoldment to humanity, especially to the intelligentsia and more advanced people. In the near future, the efforts and work of all the above groups will meet and merge into a new philosophy grounded with daily practical applications integrating the concrete and abstract sciences. This new scheme of education and learning, which we can roughly call "Aquarian Integrated Science" for lack of better terminology, will approach life, education, spirituality, business and health in a more integrated way. This new science of the future will take into account physical, vitality, emotional, mental and spiritual aspects simultaneously.

How do we study the human being in a universal way?

Let's start with the basic Human Energy System and its components. This information is also referred to in the teachings of the Ageless Wisdom, and it can be validated easily by trained clairvoyants or psychics, using their inner sight.

The different human energy bodies

Humans are made up of an energy system, which is composed of many layers of visible and invisible components and vibratory consciousness. This system can be compared to a cellular phone to make it more tangible and concrete

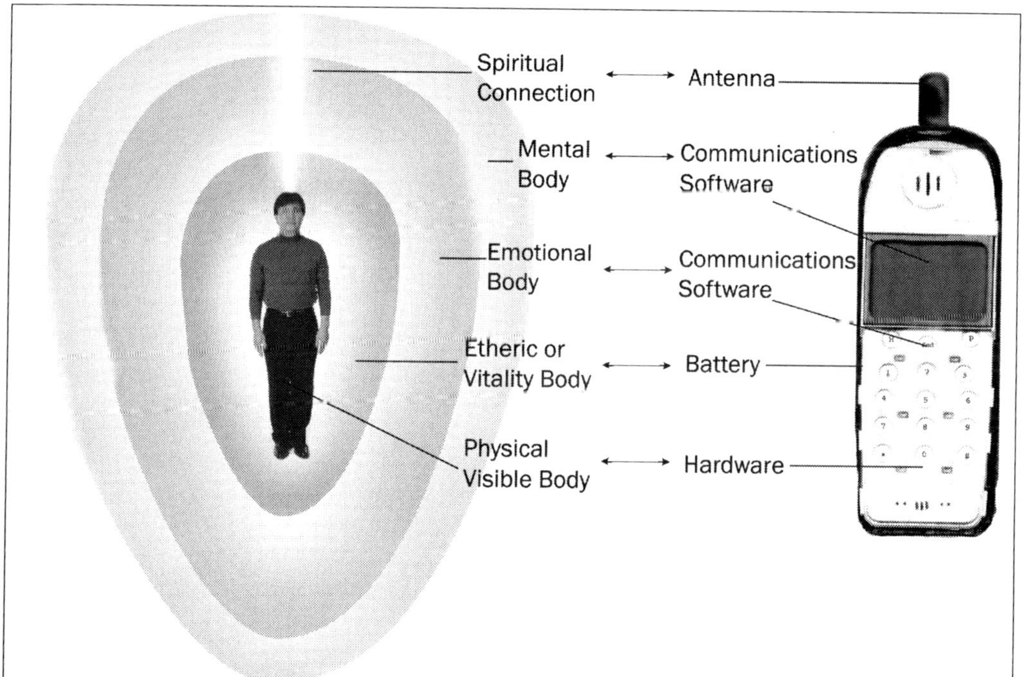

Human Energy System (Lower Bodies) **Figure 2** **Cellular Phone**

Human Energy System	Cellular Phone
Physical visible body	Hardware
Vitality or energy body	Battery
Emotions and mind	Software and programs
Spiritual connection and Divine Cable	Antenna and signal

Let's enumerate and discuss the significance of the different components of the Human Energy System:

First component of the Human Energy System: Physical body

This is what we visibly see as the physical body with a solid, liquid and gaseous construction, which the doctors have substantially studied and which almost everyone recognizes as their whole human self. This part of ourselves is the grossest and lowest vibrational aspect from an energetic standpoint. When we physically die, it gradually turns into physical earth substance. We see solids with our naked eye because they vibrate at about 16 cycles per second. Like the cellular phone, if the body gets physically sick, we bring it to the hardware specialists - the medical doctors, massage therapists, chiropractors, osteopaths, herbologists and other alternative health care practitioners.

Second component of the Human Energy System: Vitality body or etheric body

The vital body is commonly called the "etheric body" in healing and yogic literature. It acts like our battery. When you speak of vitality or stamina, this energy is the controlling factor. If we use vitality without replenishing it, we get depleted and discharged. When this vitality is constantly drained, which can be caused by stress, it accelerates the aging process or the person gets sick. As with different cell phone models, some people are born with long life and others are easily discharged and drained. This depends on their vitality, emotional, mental and spiritual development. The good news is that almost everyone can charge or increase their vitality through special methods: inner breathing, physical exercises, sleep, proper diet and meditation.

The vital body is invisible to the naked eye, but many children or psychics can see it using their inner eyes. That is, they can see the aura of the etheric body, even with closed eyes or not looking directly at you. They can see a light around a person. The etheric body has many layers of refinement. The densest part is seen as the exact shape of your body, which extends beyond the body a few inches for most people. It is bigger in more advanced and powerful human beings. This denser part of the etheric body is sometimes called "the etheric double" because its shape follows the contour of the physical body. The aura of the etheric body is like an egg shape with the tapered end pointing towards the bottom. The natural energy from the air, ground, sun and food can charge the etheric body. When you do deep breathing, the oxygen revitalizes your blood and body, but it is air vitality that charges the etheric body. The etheric body, with all its chakras and meridians (or pathways of light substance and its energy), controls, purifies and revitalizes the physical body and all its parts. The health and power of the physical body depend primarily upon the etheric body and its energy centers.

Third component of the Human Energy System: Emotional body

The esoteric literature calls this the astral body – an energy field comprised of the substance of feelings. When you feel or emote, you are utilizing the emotional substance in the emotional body. Love, joy, peace, compassion, conscience, hatred, anger, fear, irritability, jealousy, aggressiveness, sentimentality and stress are all housed or connected with the emotional body and its energy centers. People express positive or negative emotional attitudes because their emotional body comes with various emotional substances, which have been inherited from ancient lives as well as through their parents during the process of insemination or during pregnancy, a phenomenon illustrated by the fact that even newborn babies display varied qualities just after birth.

You may have experienced effects on your emotional body without knowing how to explain it. For example, sometimes when someone is angry with you, but hasn't communicated the negative feelings towards you verbally, you can feel it, especially in the stomach or solar plexus area. Likewise, when you enter a very stressed area like an office, haven't you felt your stomach knot up or get nauseous? What's happening is that you are feeling the negative vibration through your emotional body, the energy body receives and processes feelings. Another good example is mothers who are sensitive to the needs or problems of their children or partner: Even at a remote distance, they can feel their issues or needs. Why? They are connected to their loved ones through their emotional body.

You can purify the negative feelings of your emotional body and recharge it with positive, pleasant emotions using special breathing and visualization techniques. As with the etheric body, the emotional body's densest vibration is shaped following the contour of the physical body, but extends beyond the outline of its etheric body counterpart. Since the emotional body is more refined than the etheric, its vibration interpenetrates it and is more expanded.

Fourth component of the Human Energy System: Mental body

> The mind is not the brain.
> The mind is the software and the brain is the hardware.

This part of your human constitution is one of the most important because it differentiates your bodies of existence from the animal kingdom. To date, most animals use only their physical and emotional instincts while humans use their minds.

The mental body is the energy field for creating, processing, receiving and projecting thoughts. When you are thinking, which is different from feeling, you are using the substance of the mental body. As with the emotional body, most people have already experienced some of the inner powers of the mental body. For example, when someone is thinking about you - if you are sensitive and receptive at that time - you can receive the information they are thinking or at least know that the person is thinking about you and wants you to call him or her. Have you ever thought about what prompts you to suddenly have a strong urge to call someone like a family member or friend? Well, what's happening is telepathic rapport. It turns out that you call the person who was thinking about you or sometimes you end up calling each other simultaneously. Have you experienced this? This book will teach some methods to develop this inherent faculty of telepathy, which can be particularly effective when cell phones don't work in remote places!

The mental and emotional bodies serve as the software and energetic program that operate the entire personality, especially the normal mundane activities. For more advanced people, the Soul uses the software of the mind, emotions and vitality to program the personality and daily life of the individual with more noble principles and advanced inner powers and faculties. Thus the difference between a Sage and a Saint, an intellectual and an average human being. The Sage is regulated and inspired by the Divine Spirit, whereas the Saint is regulated and empowered by the Soul. The intellectual person is regulated by their mind body and partially by their emotions. The average person is functioning and active at the level of the lower emotions and physical body and is engrossed with excessive appetites.

The mental body also has many layers of frequencies. The densest can be seen psychically as the same shape as the physical body, but bigger and more expanded than its emotional and etheric counterparts. The mental body can be purified and recharged through many methods, including inner breathing techniques, affirmations and visualization. By using the faculties of the mind constantly, it is being exercised and is developing mental powers and stamina. When the mind is not used for a long period or stretched regularly with bigger ideas and new concepts, it starts to crystallize and atrophy. In a comatose person, the mind is disconnected from the brain. The mind – as distinct from the brain – will be a subject of new research and study in the coming generation.

Fifth component of the Human Energy System: Spiritual antenna or cable of light

The antenna signal of the cellular phone affects the usability of its communication equipment. The same is true with the human being. Even though the cellular phone hardware, battery and software programs are in good condition, if the antenna is not functional, communication cannot be established and the cellular phone cannot work properly. At times, the signal is distorted, so the voice is blurred or faint. This is also true spiritually for a person. The spiritual signal and communication from the Soul can be disrupted or is weak.

Many top successful CEOs and political leaders have expressed to me that they feel deep loneliness and emptiness despite having successfully accomplished most of their goals. The spiritual connection that provides the purpose of the Soul in its present incarnation to move to the next great step is what's missing for some of these successful leaders. And sometimes the Soul's will and purpose run contrary to these leaders' decisions or normal reasoning. I have found in my experience that few techniques like meditation and deeper inner reflections, either in a retreat setting or in a consultation room, will help process a few obstructing misconceptions about what is important in their lives. Usually after several sessions, the spiritual antenna starts opening and connecting, and inspiration comes into the person.

As a result, some of my clients are able to process spiritually and simplify their lifestyle and use their financial and material resources to create or support nonprofit organizations and charities supporting world needs. As soon as they are spiritually motivated, these successful leaders become the best agents of global transformation.

Many rich businesspeople and successful professionals in India and Asia, and now starting in Europe and America, leave their families, businesses and careers just to look for answers in their search for spiritual truth and the unfoldment of inner powers in order to make the next great leap of their lives. Greatness normally demands great sacrifice, selflessness and total surrender. In my own quest walking and climbing in the Himalayan Mountains, I came across renunciates, who used to be wealthy and very successful executives from Europe, North and South America and Australia, dressed like beggars with long beards and sandals. When I asked them why they left their homes and sacrificed their material and social life and families, their answers were all similar: to search for a different, fuller meaning to life and existence.

I believe that we do not have to drop everything, including our past achievements and loved ones, in this search. A spiritual science offers modern techniques to discover the light of fuller existence and service to achieve total fulfillment without abandoning our normal lives. The three volumes of *Inner Powers* offer such technology for spiritual advancement. In order to integrate and explain the more advanced principles and techniques of inner power development properly, teachings on the nature of the Soul and Spirit have to be emphasized.

Higher component of a Human Being: The Soul

A human being is not a physical body with a Soul and a Spirit. It is indeed reversed. A human being is a Spirit using the Soul to incarnate through the personality to implement goals that allow it to become the master of life. The personality of the human being gets its life and power supply of electricity from the Soul who gets its supply of power from the Divine Spirit – the "I am that I am" in every person.

How is the brain connected to the mind and the mind to the Soul and the Soul to the Spirit and the Spirit to the Macrocosm we call God?

This is a common question, even among advanced students. The incarnated person has three energy bodies that control the human energy system: the concrete mental, emotional and etheric or vitality. The Soul has different connections to all of these energy systems:

- The center of the head, approximately at the pineal gland, is the anchor of the mental light connecting the brain, the concrete mind and the Soul. This is located at the mental frequency.
- The center of the chest along the body's central axis approximately at the sino-atrial node of the physical heart is the anchor of physical and etheric light connecting the etheric body with the Soul. This is located at the etheric frequency.
- The center at the solar plexus area below the sternum near the tip of the liver is the anchor of the emotional light connecting the emotional body with the Soul. This is located at the emotional frequency.

All of the different energy bodies have connections and correspondences at the Soul levels. The Soul as a being also has energy centers, which are qualified with spiritual will-power, love-wisdom and active

Anchor Points of the Soul in the Different Bodies

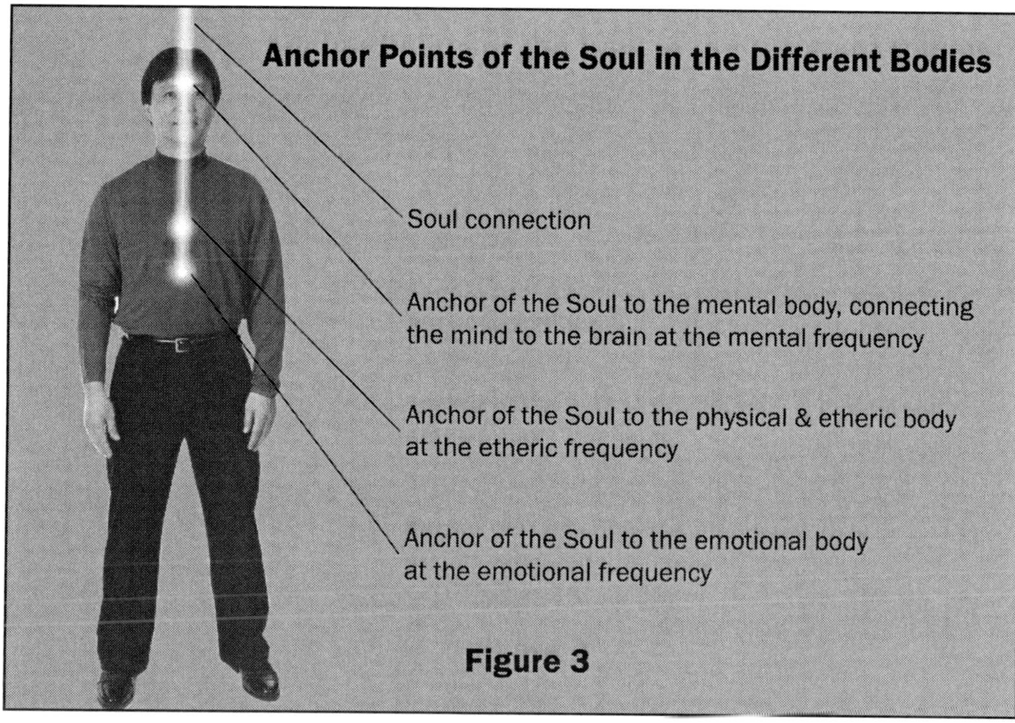

Figure 3

intelligence. The Soul also has equivalent connections to the Spirit or Divine Self. The Spirit is connected to the main power supply, which is the Planetary Divine Macrocosm, the Source of Divine Life and Light. In the English language, we call him/her God, but before the English language was created, this Planetary God was called differently by many ancient religions.

This constitution explains the locations of the connection of a human being to the Planetary Source of Life and Power. Let's look at a human body like an electrical system to see how all of this works:

- The lighted electrical lamp is the personality
- The transformer station is the Soul

- The major power plant and source of Divine electricity is the Spirit

When the Soul wants to die, it pulls its electrical connection from the physical and vitality bodies. Therefore, just like the electrical lamp, its power supply is switched off.

The real source of inner power is the Soul and Spirit, which are empowered by the Planetary Macrocosm. The ability to channel, absorb, assimilate and use the Divine Light and Power is the real stimulant for inner power, which leads to greater service and usefulness in life. All the great achievements utilizing abstract and concrete intelligence and the ability to concretize great ideas, projects and plans are by-products of the stimulation and empowerment brought by the Soul or the Spirit to the incarnated person.

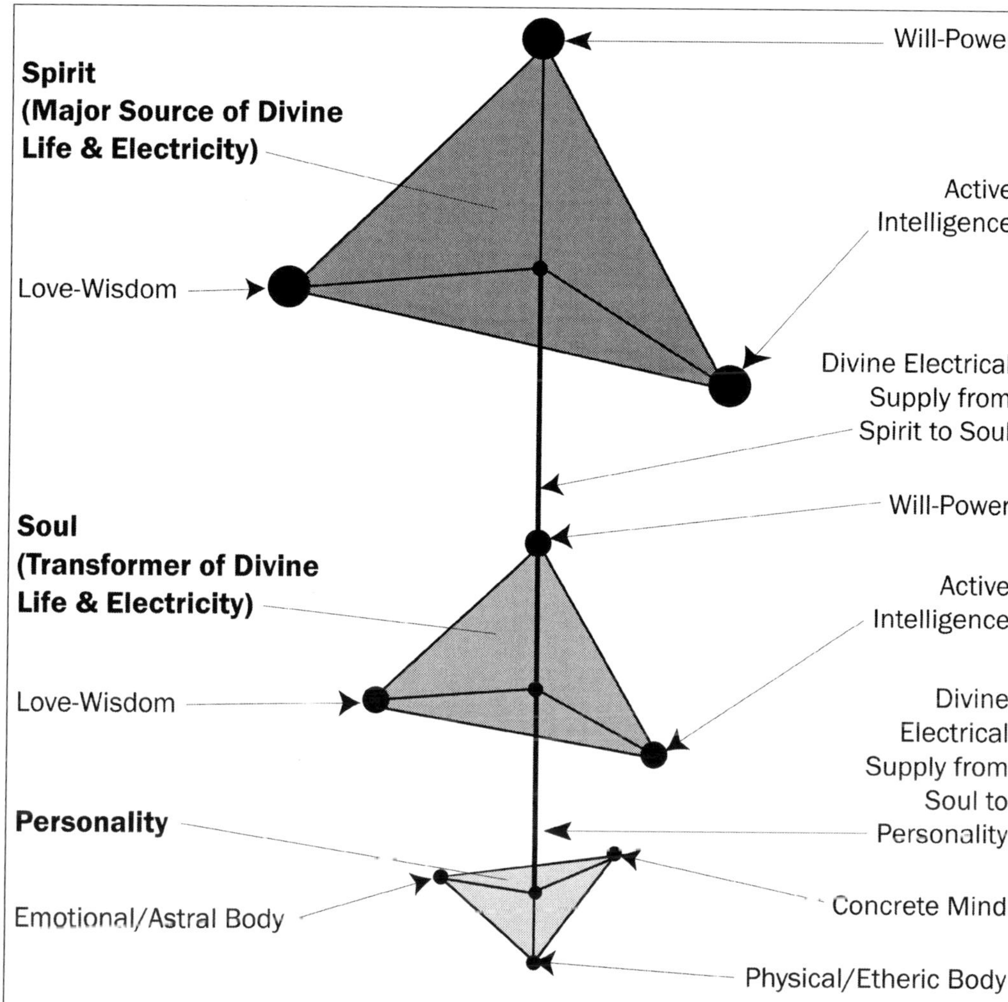

Schematic Diagram of the Spirit-Soul-Personality Relationship
Figure 4

Highest component of a Human Being: The Spirit

The Spirit is the Divine Self or the God within. It is the "I am that I am" within every person, which allows him or her to become a true Spiritual human being. It has three Divine qualities: will-power, love-wisdom and active intelligence, copies from the Divine Macrocosm. This is what is popularly known in Christianity as one God and three Divine personifications, the Trinity. The Spirit or Divine Self is the giver of life and light to the Soul. This subject is discussed in more detail in Section E on Spiritual Powers.

Human energy centers or chakras

There are many chakras or energy centers in your system. The word chakra is a Sanskrit term meaning "whirling wheel." In this current cycle, these energy centers in humans look like wheels or vortices of invisible light that form petal shapes anchored in their center. Their roots stem out from an axis of light crossing the center of the body. Most of the bigger chakras extend horizontally to the front and back of the body, and a few unfold upward and downward.

Sample Chakras - 3 Layers
Figure 5

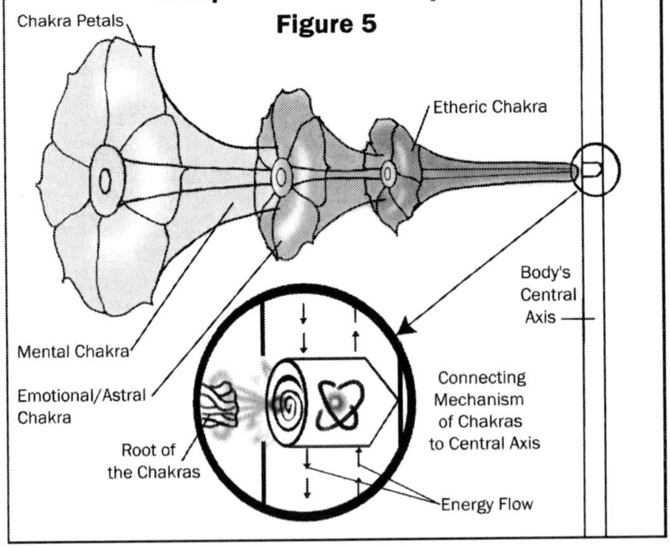

Functions and descriptions of the energy centers
1. The chakras are centers of energy that purify and revitalize the organs and systems of the physical body.
2. These centers are entry points, assimilators and processors of energy needed by the bodily functions.
3. The energy centers act like software or a subconscious program for the proper functioning of the bodily organs and systems.
4. They serve as a security gate for unwanted vibrations and negative forces, preventing them from entering the bodies.
5. They serve as a transmitter of energy and information within the bodies, into and out of the energy bodies.
6. These chakras are also centers for psychic and spiritual faculties and inner powers. These faculties will be tabulated in Sections D and E.
7. Some energy centers are utilized by the Soul to control certain physiological and psychological functions, including karmic executions and dispensations of what we call good and bad situations in life.

Current forms and features of the chakras in human beings

The chakras have different stages of development according to the evolutionary level of the human being. As the person becomes a Saint or a Sage, a transfiguration of the total Human Energy System and the chakras occurs. Therefore, the shapes and features change. Here are general descriptions for the introduction of this topic:

- Chakras are formed by the aggregate crisscrossing of invisible forces and energy fields in the different energy bodies. Up to the current cycle of the Earth, humanity's chakras are circular and are anchored in the central axis of the body like funnels or vortices of light.

- The movement of the chakras is clockwise when they are admitting energy and counter-clockwise when they are releasing energy. The continuous rotation back and forth with all the interacting energies and forces creates spokes or petals, similar to the effect of a kaleidoscope. The faster the rotation of the chakras, the bigger the vortex it creates, and this factor determines the size of the chakra.
- The chakras have different colors, which are vibrations that stimulate certain physiological functions of the organs and also trigger the person's psychological behavior.
- There are three layers or tiers of chakras starting at the etheric frequency and continuing into the emotional and mental bodies.

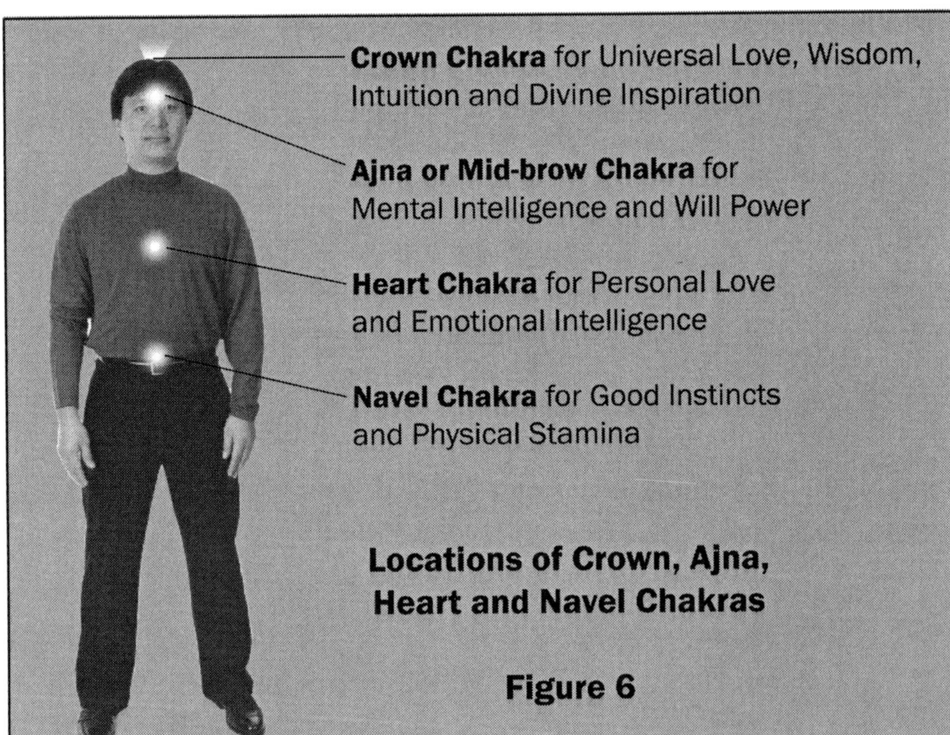

Locations of Crown, Ajna, Heart and Navel Chakras

Figure 6

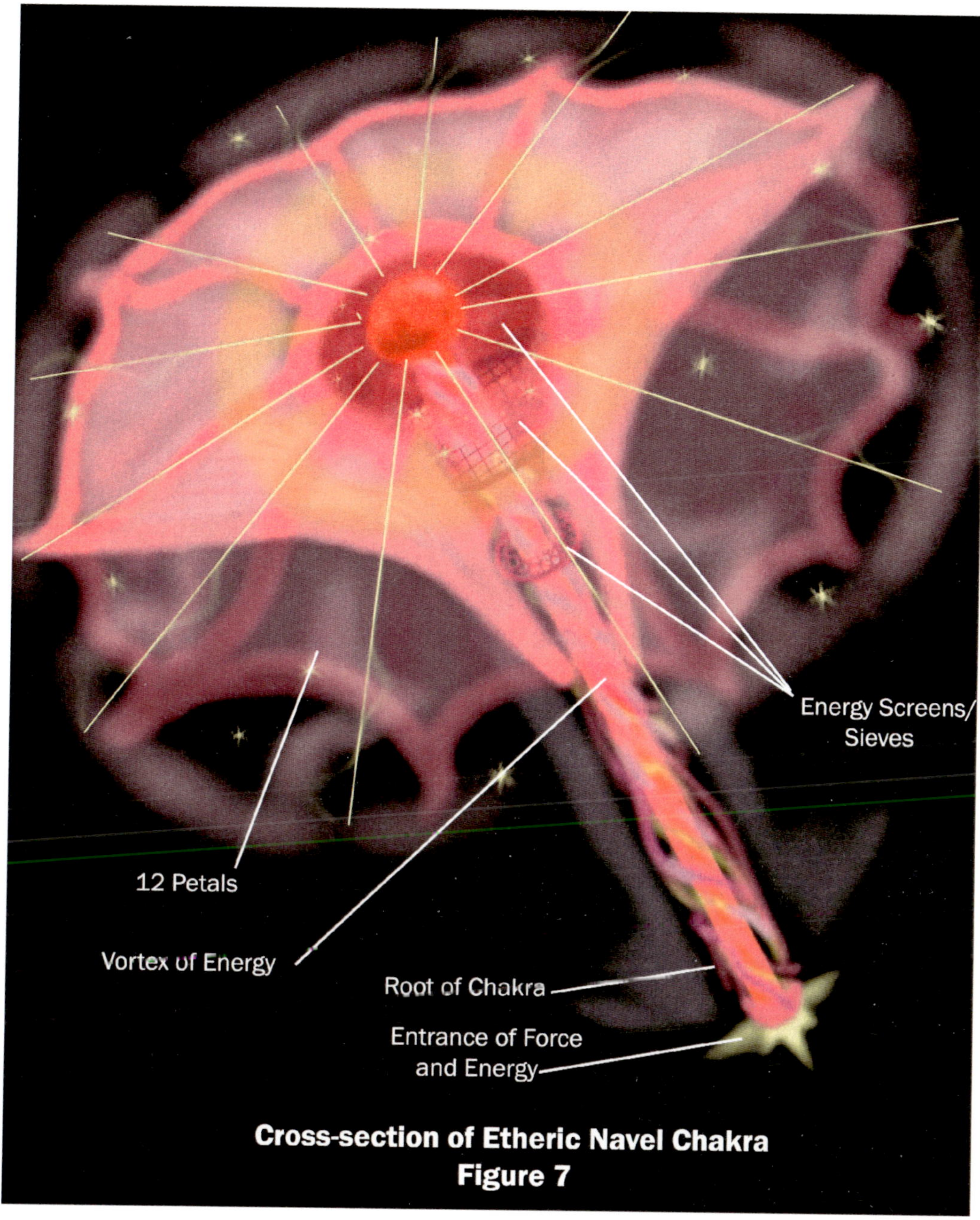

Cross-section of Etheric Navel Chakra
Figure 7

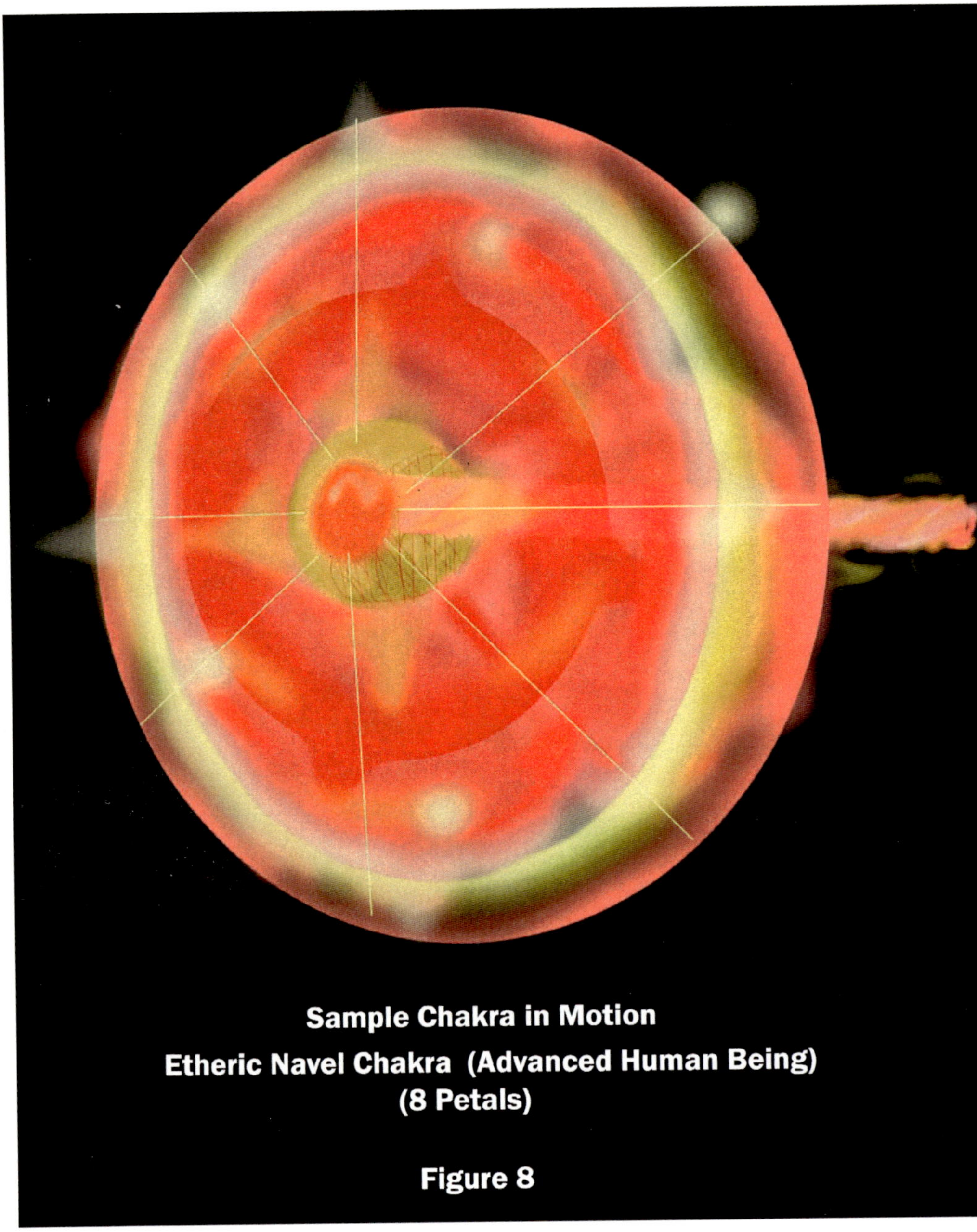

Sample Chakra in Motion

Etheric Navel Chakra (Advanced Human Being)
(8 Petals)

Figure 8

There are three sizes of chakras: the major, minor and mini-chakras. Most of the energy centers of the front and back of the body anchored along the central axis are the bigger chakras. The organs have smaller sizes of chakras, and some part of the body like the fingers have mini-chakras.

The higher the evolutionary development of the person, the bigger their chakras. The following is a list approximating the etheric chakral size according to level of development:

Approximate Chakra Sizes of Human Beings	
Early humans	2 inches or less/5.08 cm
Average humans	3-4 inches/7.62-10.16 cm
Intelligentsia	5-8 inches/12.7-20.32 cm
Advanced humans	9-15 inches/22.86-38.1 cm
Saints	16-24 inches/40.64-60.96 cm
Sages	25 inches or more/63.5+ cm

In most cases, a person's chakras are not perfectly balanced or aligned (i.e., some of their chakras are bigger than others). Also, within the range of each of the above groupings, there are many sub-levels of development, thus the difficulty of approximating chakral size.

Chakras for Initial Study of Inner Power Development

Chakras for Initial Study of Inner Power Development
Crown Chakra
Mid-Brow or Ajna Chakra
Heart Chakra
Navel Chakra

To initiate you into the basics of inner power development, we will first focus on a few chakras. There are four important chakras to initially develop vitality, emotional and mental intelligence and spiritual powers, the faculties stimulating the powers of wisdom and intuition. These centers are very developed in Saints, Sages and advanced human beings. That is why some of them do not have formal educations, yet their ancient works of art and philosophy are masterpieces to our present thinkers.

Let's start with the centers that are needed for the practices of the next sections. The other chakras will be studied in a later section and their technical descriptions and colors will be discussed in Volume II of *Inner Powers*.

First chakra for inner power development: Crown Chakra

This is a center or vortex of energy called "crown chakra" in yogic philosophy. It is located at the top of the head. It is one of the few chakras that sprouts along the axis of the energy bodies. It looks like a funnel standing and twirling on top of the head. The teachings of the Ageless Wisdom describe the crown as having 12 inner petals and 960 outer petals[1]. This can be seen by trained clairvoyants. When this center is stimulated or activated to a high degree – especially for a long time – it harnesses the powers of intuition and universality and brings down wisdom and inspiration. That is why the symbol of wisdom or spiritual leadership and enlightenment is the crown and the hats of different Saints and Sages:

- The cone shape on the head of the Buddha, which symbolizes full illumination and wisdom with a complete set of 1,000 all-seeing eyes equivalent to complete enlightenment (Figure 9, page 36). The Buddha is said to be the embodiment of full wisdom and enlightenment.
- The crown of Jesus the Christ with 12 stars and points (Figure 9, page 36). When seen with inner sight, this power center has 12 golden petal-like energies and when fully developed, it has 12

opened petals like a crown or opened lotus flower. The Christ is a manifestation of a completely perfected human being.

- The symbol of holiness and office of spiritual leadership in the Christian religion is a mitre worn on the head by the Pope and bishops. This crown symbol is also embodied in the Christian teachings of the 3 kings who visited Jesus the Christ when he incarnated in Bethlehem.
- The Native American shaman's many-feathered headdress, which symbolizes wisdom and spiritual leadership as the tribal head
- The headdress worn by the Egyptian pharaohs, which symbolized power and wisdom in leadership (Figure 10, page 38)
- The round halo of saints and Christian angels, which is the artist's portrayal of the crown center.

As you can see from these examples, the crown chakra or energy center has universal application and acceptance, but only those who open their minds and question these spiritual symbols will most likely understand their deeper meanings. The crown energy center is one of the most important power centers to be understood, activated and empowered for spiritual faculties and higher inner power development.

The Crown & Heart of Mary, the Mother of Jesus

The Crown of Jesus

The Crown of Kings & Queens

The Crown of the Buddha

Figure 9

Second chakra for inner power development: Mid-Brow or Ajna Chakra

This is another power center that can be seen externally in some cultures. An example is the Indian women who, when they mature, put a red dot between the eyebrows. Most yogis emphasize stimulating this point, as do the internal martial artists and esoteric acupuncturists. This center activates mental will-power as well as abstract intelligence. Have you observed that when you need to push heavy objects and boost your power, you squint your mid-brow area naturally? It tends to boost your will-power. Also, as you work on penetrating and concentrating on a certain deep thought, you tend to furrow or squint this area, which increases your ability to focus. If you observe people who are willful, they have more wrinkles in the forehead area and between the eyes. Great leaders usually employ great will-power, vision and abstract thinking using the mid-brow center, even unconsciously. The higher integration of this power center with the other chakras in the head is symbolized by the snake's head at the mid-brow of the pharaohs, the symbol of the inner eye or Third Eye.

Third chakra for inner power development: Heart Chakra

The center of power that is probably most familiar to people is the heart, which is analogous to emotional intelligence and the love nature. One of the most powerful instruments to be able to sustain leadership is the ability to be liked by people. The faculties of love and compassion, altruism and benevolence utilize the heart center. This is commonly depicted in portraits of Saints of different religions as the brightness and beam of golden light emanating from the center of their chests.

The Heart & Alta Major Centers (at the back of the head) of Jesus and His Mother

The Cobra & Volture's Head in front of the Pharaoh's Headdress, Related to the Ajna Chakra and a Symbol of the Inner Eye

Figure 10

Do you know why saints and gurus are characterized as "big-hearted" and as "having a heart of gold"? When the heart chakra is highly activated, it has a pinkish and golden color, especially in people who practice compassion, generosity and loving-kindness. It is an invisible vortex of energy with 12 energy petals at the mid-chest area. That is why when you fall in love or have done good things for others, you feel a pleasurable, warm sensation in your chest area. Likewise, when a person violates the heart's virtues against others, the chest feels heavy and shrinking, and it feels closed. This center of light expands or shrinks depending on the activity and emotions executed or received by the individual. By activating this power center, you can feel love, peace and happiness, and the ability to forgive others becomes natural.

Fourth chakra for inner power development: Navel Chakra

This center of power is most used by martial artists and advanced yogis in their inner practices and meditation. By breathing slowly and focusing on this power center, you can improve the following:

1. Internal power for stamina and the body's rejuvenation
2. Absorption, assimilation and circulation of internal vitality as a reserve of power
3. Instinct and agility in your physical reaction and movements
4. Speed, instincts and timing of decision-making
5. Sexual potency

This energy center is weak and underdeveloped in many people who move slowly, have poor timing in decision-making, are easily devitalized, or are sexually impotent.

The navel chakra is an ancient secret of great stamina and instincts in martial arts and boxing champions. It is one of the most important energetic developments separating the super-athlete from the ordinary. If you look at the best golf and basketball players psychically or clairvoyantly, they have one thing in common: Their navel energy centers are more active and developed than in most players.

In martial arts, there are two kinds of practitioners. The normal type, using more brute or physical force, is the exoteric practitioner. Those who use internal energy, including some of the more advanced *chi-gong* and esoteric *tai chi* methods, are the really powerful martial artists. These more advanced practices involve breathing and focusing on the navel area to activate the navel chakra and its energy reservoirs. Focusing on the navel charka will give you "inner centeredness" and very good instinctive awareness.

The faculties of the navel center apply in the business world as well. When a clairvoyant or a good psychic examines the energy and auras of the best business decision-makers and great leaders, they can see that the navel energy centers are more developed and fully charged than their subordinates or peers.

Why then do these decision-makers sometimes commit big mistakes or decisions? Because there are different degrees of instinctive knowing, and this is just one of them. Sometimes higher discrimination is needed to override instinctive knowing. These higher faculties are developed from the crown, ajna and heart centers. It is not only the development of the navel power center that is important, but the amount of energy available in it. When the vitality of the navel chakra is low, instincts are off. Therefore, there is a need to constantly activate and energize the navel energy center.

I have experimented with activating and energizing the navel chakras of players on some sport teams, and I have found that it definitely improves their performance immediately. When I climb mountains, I boost my internal strength and stamina using navel center breathing methods that improve my mileage dramatically, especially in high altitude without having to acclimatize. I can imagine that if these techniques were taught to astronauts, their health would improve and anti-gravity negative side effects like the usual muscular or skeletal deterioration would be minimized.

This navel breathing technique has been taught to sexually impotent clients and after several weeks, their problems were solved. In the future, it will be a must for Olympic and professional athletes to utilize this technique to break records in their chosen fields.

The ability to systematically and esoterically manipulate the energy centers corresponding to certain faculties is one of the secrets of the Sages in developing inner powers. Most of their techniques to boost vitality rapidly, develop a super-mind and precise conscience and embody a compassionate nature are introduced in the following sections of the book. So let's go practice the different techniques to maximize our performance and improve the quality of our life.

Part II
Inner Powers & Their Types

Section A: **Power of Vitality: Principles & Techniques**............... page 43
Sources of the power of vitality.................................... page 44
Special exercises & breathing techniques
to build vitality.. page 51

Section B: **Emotional Powers & Intelligence:**
Principles & Techniques... page 89
Ramifications of emotional intelligence...................... page 90
Techniques to develop emotional
power & intelligence... page 97

Section C: **Mental Power: Principles & Techniques**................. page 115
Power to physicalize... page 119
Mental experiments & exercises................................ page 120
Procedures for programming & materializing goals... page 125
Ramifications & elements of mental power................ page 130
How the mind works... page 138
Practical techniques for mental power
& its applications.. page 164

Section D: **Psychic Powers: Principles & Techniques**................ page 179
Psychic power experiments... page 182
Attitudes for psychic power development.................. page 185
Different types of psychic powers............................... page 194
Basic requirements for psychic
power development... page 195
Practices for psychic power development.................. page 208

Section E: **Spiritual Powers: Principles & Techniques**.............. page 287
Questions about spiritual powers................................ page 290
Triplicity in a human being.. page 297
Major sources of spiritual powers................................ page 309
Methods to develop spiritual powers........................... page 331
Virtues for spiritual inner power development............ page 338

Section A
Power of Vitality: Principles & Techniques

I. Sources of the power of vitality....................page 44

A. Earth's natural energy....................page 46
B. Energized people....................page 50
C. Invisible forces....................page 50

II. Special exercises & breathing techniques to build vitality.... page 51

A. MDP Body Therapy & Inner Power Development™ Program ....page 51
B. The science of breathing for inner power development.... page 72
C. Slow Abdominal Breathing Technique (1st type)....................page 77
D. Slow Abdominal Breathing Technique (2nd type)....................page 78
E. Combination Breathing Technique....................page 81
F. MDP Synchronized Breathing Technique™....................page 81
G. Weekly vitality management schedule....................page 87

Part II
Inner Powers & Their Types

Just as there are different expressions, qualities and degrees of business acumen, so there are different expressions, qualities and degrees of inner powers. The influence and permanence of the effects of inner powers are based on how successfully the inner faculties and powers are nurtured, how they are applied, what force and energy are supporting them and how integrated the inner powers are when applied for greater service and group work. When the spiritual principle of the Soul or Spirit has been anchored in the personality, expressing itself through qualities like benevolence, altruism, discipline or group consciousness, the person already has substantial inner development and has developed the habit of doing the right thing. Thus, the use of inner powers is maximized for higher purposes.

Section A
Power of Vitality:
Principles & Techniques

Why is vitality critical to inner power development? Have you ever experienced waking up in the morning feeling that you are too weak or too old to tackle big things or to start new projects? Or sometimes it feels as if power has gone out of your life, and you fear that your life won't last long? Why does this happen? It has to do with will-power or vitality. If your vitality is low, you are like a cell phone or car with a dead battery. You can't function properly. No matter how intelligent, determined or enthusiastic you are, when your energy and will-power are down, you can't function well. Vitality is like your car battery and the gasoline fueling your accomplishments and success.

Without it, your talent and ambitions don't work. Many great successes fail when vitality is drained.

Vitality levels also affect mental and emotional aptitude and attitude. In my coaching services for leaders, executives and professionals, one of the single most important factors for them to be on top of their life and achieve goals is the ability to maintain constant vitality. Better decision-making, problem-solving, negotiating power and quality relationships are possible with good vitality. Maximizing your performance depends on overcoming stress and critical difficulties in life, which in turn depend on your vitality. Even your ability to overcome fears, depression and diseases depends on your ability to maintain vitality. For this reason, I am devoting a whole section of this book to vitality development and management. One of the basic foundations of this maximizing performance work is my MDP Body Therapy & Inner Power Development Program, which gives you fast methods and techniques to be able to maintain higher vitality and balance your life daily. We'll discuss the techniques later in this section.

Sources of the power of vitality

Sources of the Power of Vitality
Earth's natural energy: Sun energy Air vitality Earth power Natural storage areas
Energized people
Invisible forces

Many varied sources of vitality and inner power exist. The Earth's natural energy is one of them. It brings the power of vitality and restores health via:

First source of Earth's natural energy: Sun energy

Based on my studies with my early spiritual mentors, who all share a basic principle involving the source of Earth energy, the foremost source is the sun. There is a force from the sun that gives power to the air, the physical Earth, water and the auras of the visible and invisible parts of the Earth. To receive some of the sun energy force, it is best to bathe in the sun's energy for a few minutes in the morning. Medical experts also say that the sun's energy provides vitamin D, which facilitates the absorption of vitamin E in the body.

In my healing practice in India and the Philippines, I experimented with having sickly or weak individuals bathe in water exposed to the sun for at least one hour. I have observed that this can energize a person's body and vitality, especially for old and weak patients. This method energizes the bodies and potently detoxifies the body's energy if salt is diluted in the solarized water. Any salt works, but sea salt is more natural. This salt and water solution, mixed with 1 cup of salt per 2 gallons/7.57 liters of water and empowered by the sun, gives a purifying, energizing and healing effect. This is supported by the many sick people who travel to the Dead Sea to experience that potent salt water's curing effect.

Second source of Earth's natural energy: Air vitality

The air you breathe is not only a combination of oxygen, nitrogen, other gases and toxins; air energy vitality also exists, which is occasionally visible to the naked eye. Scientists sometimes interchange this with the ions in the air. The air is very congested during a thunderstorm because of the predominant positive ions, but after the rain, it is refreshing because there are more negative ions in the air. This ionic "air power" is called "air globules" or "air prana" in the Ageless Wisdom.

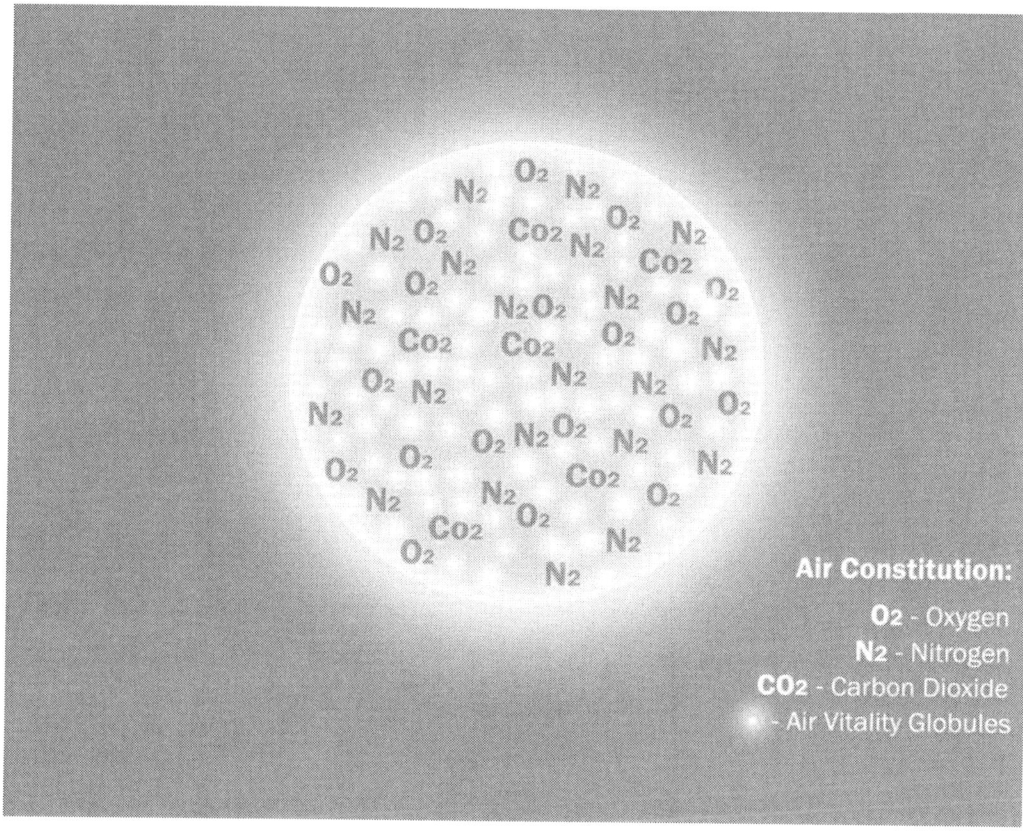

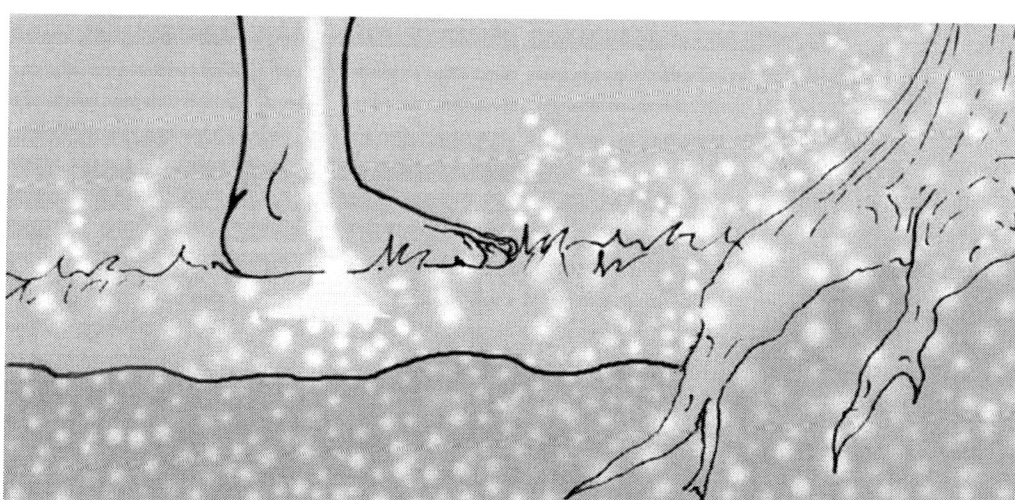

Earth Energy Being Absorbed by the Sole of the Foot

Figure 11

You can sometimes see these globules moving in the air sporadically. Look at the blue sky for a few minutes and just be aware of the air until you see something in front of you…like points of light. These are the air vitality globules, which are so essential for human life to exist. At night, especially after midnight, the air's energy is less potent due to the absence of sun energy. The air gets its vitality energy from the solar rays. Therefore, it's good to expose spaces inside your home to sunshine and fresh air to revitalize its energy, especially in basements and dark spaces.

These sources of inner power are already common knowledge to healers, yogis and internal martial arts practitioners. In the future, medical doctors, scientists and sports coaches will also augment their methodologies with this additional air and sun power in mind. The manipulation of the body through massage therapy and rejuvenation through very physical forceful methods will increase its effectiveness by adding a no-touch healing method using these additional energies. *Tai chi* and yoga practitioners currently use many of these techniques, but many have not yet maximized their full potential.

Third source of Earth's natural energy: Earth power

Some tribes in Africa and Asia are known for healing sick people by putting the patient's body in a hole in the ground or in caves. They use different types of holes for purifying and for vitalization. When people faint because of depletion, one of the fastest ways to recover their energy in the absence of a medical doctor is to lie them down on clean ground and just let them relax.

The Earth, having a natural power absorbed from the sun as well as latent power in itself, radiates a powerful, dense energy and vitality that you can also absorb through your feet, especially if you are barefoot. By consistently absorbing the Earth energy through the feet using slow breathing and deliberate intention to bring it to the body, you can assimilate this power to tone the muscles, strengthen the bones and rejuvenate the body. Lying on the beach for an hour is also a good energizing technique. You can augment this practice by swimming in sea water, then exercising. This combination of a sea water bath and exercise, plus lying in the sand, is a great detoxifying and revitalizing experience. I usually advise smokers and recovering clients to do this for fast revitalization and healing.

Fourth source of Earth's natural energy: Natural elements in water and food are sources of vitality and energy for the physical body

Proper diet and nutrition are an essential part of vitality management. Healthy food, good sleep, proper physical exercises and an abundant supply of oxygen are a great combination to maintain your vitality. In terms of diet, many advanced spiritual practitioners prefer vegetarianism or semi-vegetarianism diets to maximize the effects of their inner practices. They have observed that the energy vibrations of vegetables are higher and more refined than the vibrations of meat.

Fifth source of Earth's natural energy: Natural storage areas

There are many special places on Earth that have higher or more positive energy. By being inside these so-called energized places, you get energized, revitalized and empowered rapidly. There are also highly healing and vitalizing locations like areas where mountains, rivers or springs and oceans meet. These areas accumulate negative ions for healing energy. Meditation centers, temples, churches and even caves where Sages have experienced their enlightenment have strong and refined energies. Areas or plains with a mountain in the south or west are endowed with positive energy.

Source of the power of vitality: Powerful individuals tend to energize people & groups

Energized people tend to be a dynamic and enthusiastic source of vitality and power. Powerful healers, martial artists and yogis who practice internal energy breathing techniques and exercises are able to transfer this vitality to others easily. There are also many powerful executives and athletes who can transfer their dynamic energy to others just by being around them. Energy flows like water from the higher level to the lower. Spiritual teachers, Saints and Sages are a tremendous source of energy. Spiritual groups are also a depository of spiritual substances, whereas wealthy individuals and groups have great material energies of prosperity. The nature and quality of people and groups determine the type of power and energy they possess.

Source of the power of vitality: Invisible forces

There are many invisible beings who are depositories of vitality for different purposes like healing or transformational requirements of humanity. In Sanskrit, they are called *Devas*. They are called angels in Christianity. These beings are also used for different services like vitalizing thoughts and ideas of many groups. They have different grades of development and belong to group consciousness rather than being individualized like humans. Their colors depend upon their work and their group. When people are doing rituals, ceremonies, meditation, chanting, healing or are performing their service for a good cause, these invisible beings and helpers are attracted to help. When you speak a word or think, it is a constructed form to them.

Special exercises and internal breathing techniques to build vitality

MDP Body Therapy & Inner Power Development Program

The MDP Body Therapy & Inner Power Development Program offers special techniques to develop inner power and internal energy. I used to exercise two hours a day in order to stay fit during my martial arts days, especially before competitions and tournaments. I used to employ physical training power more than internal power. As I learned the secrets of internal energy through my healing mastery, I combined both of these methods and designed a quick but effective way of purifying, revitalizing and balancing the human energy system. Twenty to thirty minutes of this method gives more internal energy and balance than my normal athletic and martial arts training. The program was designed with the following expected results:

- Rejuvenate the body and release stress and body tension
- More oxygen in the blood and the body to enhance the metabolism, accelerate weight loss and purify and revitalize the whole body
- Improve body strength and stamina
- Increase vitality and improve its proper circulation throughout the entire body
- Release toxins and stress from all the internal organs and the energy body
- Activate and align certain power centers of the body
- Special overall alignment of the body, emotions, mind and spiritual powers

Physical exercises and body movement alone are not sufficient to develop internal powers. The muscles can be toned and strong but lack inner stamina and internal energy. Certain additional exercises and physical biomechanical movements are necessary to release blocked stale energy and stress in the meridians and muscles. In culturing inner power, it is important not only to increase the energy level, but it is also essential to circulate oxygen through the blood in the whole body effectively and to distribute the internal energy throughout the human energy system properly. I believe that you don't have to over-exert your body to accumulate energy faster. In my experience, symmetrical body exercises with synchronized full breathing are unsurpassed in the absorption and circulation of oxygen and assimilation of vitality and power in the body systems while also accomplishing the purifying process in your aura.

During a seminar I conducted in Lausanne, Switzerland, one of my students – an 86-year-old woman – practiced these exercises and techniques. After just two sets, she claimed to feel stronger and that the pain in her body was gone. Her vitality increased dramatically and so did her mental sharpness. And when I taught the "Maximizing Performance Through Inner Power Development" seminar in Montevideo, Uruguay, an 83-year-old gentleman learned the Body Therapy and Inner Power Development Program. After the class, he stood up and declared that he felt stronger and more alive than he had for the past 20 years.

This exercise and breathing program was designed as a method integrating my martial arts and mastery of healing and spiritual sciences, bearing in mind that most people only have 15-30 minutes to exercise daily. This method does not require special equipment or accessories and can be done anywhere, anytime. You can do the exercises and breathing dressed in your business suit in your office, at a conference or during work breaks. A video demonstrates and explains all the techniques but I would like to discuss the details here:

Part 1 of the Program offers simple physical exercises synchronized with full breathing. The exercises will rejuvenate and detoxify all parts of the body from head to toe without making you tired. The exercises also oxygenate the blood and cells and enhance metabolic processes.

Part 2 provides advanced yet simple inner breathing techniques to activate and balance the inner powers of the chakras and energy bodies. Clients who have been practicing the Inner Power Development techniques report not only experiencing internal power development but also balanced love and heightened creativity.

MDP Body Therapy and Inner Power Development Program Procedure: Part 1

Body Therapy Techniques

Preparation:

1. You may follow the video guiding this program.
2. You can exercise almost anywhere with no pollutants, but best results are achieved when there is clean fresh air supply in your exercise space. I recommend that once you have familiarized yourself with the sequence, if possible, do them outdoors without following the video program.

Part 1: Body Therapy Techniques

1. **Body shake** to loosen up the entire body (2 sets): Place your feet at shoulder width. Shake your whole body to loosen up and build up the rhythm. After 1 minute of shaking, slow down and stop. Bend your knees slightly. Concentrate on the top of your head and soles of your feet simultaneously and breathe slowly. This is to allow the circulation of energy in your whole body (30 seconds). Repeat the whole Body Shake procedure one more time.
2. **Eye Therapy** to enhance your eyesight (10 times each movement): Move your eyes up and down. Move your eyes from side to side. Move your eyes diagonally from the upper right to the lower left. Move your eyes diagonally from the upper left to the lower right. Rotate your eyes counterclockwise. Reverse the rotation. Massage your eyes and face gently.
3. **Neck Therapy** to release tension in your throat and neck area. This also revitalizes the brain. (10 times each step, except neck rotations): Tilt your head slowly from side to side. Next, turn your head from right to left. Then, inhale rapidly through your nose as you move your head up and exhale through your mouth vigorously as you move your head down. Next, rotate your head slowly in either direction 3 times. Then reverse direction and rotate your head 3 times.
4. **Shoulder and Spine Stress Release** to release tension in your upper body and spine (10 times each step): Place your hands together in front of your navel. Inhale rapidly as you pull your bent arms towards the back with

your head tilted backwards and your back arched. Then exhale rapidly as your curl yourself forward, chin in and your hands back to the front.

5. **Arms and Hands T-Exercise** to empower the shoulders, arms, hands and fingers (10 times): Extend your arms together in front of your chest. Inhale fully as you move your arms to the side, with your whole body and arms in a T-position. Hold your breath and rapidly close and open your hands 3 times. Then rotate your hands around the wrists 3 times. Exhale fully as you return your arms to the front position.
6. **Arm Swing** to oxygenate the lungs and blood. This also empowers the upper torso (10 times each step): Extend your arms together in front of your chest. Inhale fully as you swing your arms around your shoulders by circling them up backwards. Exhale fully as you bring them back down to the front.
7. **Butterfly Stroke** to further revitalize the lungs and blood and empower the upper body (10 times each step): Extend your arms together in front of your chest. Inhale fully as you circle your arms back and downwards around your shoulders. Exhale fully as you bring them back down to the front.
8. **Upper Body Twists** to release tension in the upper torso and spine (10 times in each direction): Place your feet at shoulder width. Join your knuckles in front of your chest. Inhale fully as you twist to the far right and exhale as you swing back to the left. Do this 10 times, then reverse the direction and breathing for another 10 sets.

9. **Hip Rolls** to loosen up the hips and flow vitality throughout the body (10 times in each direction): Put your hands on your hips. Rotate your hips to the right in a circular motion 10 times. Then reverse direction 10 times.
10. **Internal Organs Exercise** to detoxify, revitalize and exercise the internal organs (not recommended for people with internal organ problems and pregnant women) (10 times): Place your feet at shoulder width and bend your knees slightly. Inhale rapidly while expanding the abdomen. Then exhale rapidly while compressing your abdomen.
11. **Expanding Squats** for anti-aging and increased will-power. This exercise also enhances blood and vitality circulation throughout the body (not recommended for people with knee problems) (20-30 times): Place your feet at shoulder width and put your palms together at the center of your chest with fingers pointing upwards. Inhale fully. Then exhale fully as you squat down, bending your knees and expanding your arms fully stretched to the side. Inhale fully as you come up, bringing your palms back to the center position. After completing all the squats, stand steadily. Concentrate on the top of your head and soles of your feet simultaneously. Breathe slowly and relax for 30 seconds. This technique facilitates the flow of vitality and power in your entire body.

12. **Whole Body Stretch** to align the entire body and release the pressure on the spine. This also strengthens the lower legs and feet (10 times): Place your feet together and put your palms together at the center of your chest with your fingers pointing upwards. Inhale fully as you pull your hands up as high as you can over the top of your head, stretching your whole body on tip-toe. Then exhale fully as you return to the starting position with your hands back at chest level.
13. **Side-to-Side Body Stretch** to loosen up the sides of the body and further oxygenate the lungs and blood (10 times): Place your feet at shoulder width and put your palms together at the center of your chest with your fingers pointing upward. Inhale as you pull your hands over your head as high as you can. While holding your breath, tilt your upper body to the right, then to the left and back to the center starting position. Exhale as you bring your hands down to the original position at chest level.

MDP Body Therapy Exercises

Figure 12-1: Body Shakes

(2 sets - 2 minutes)

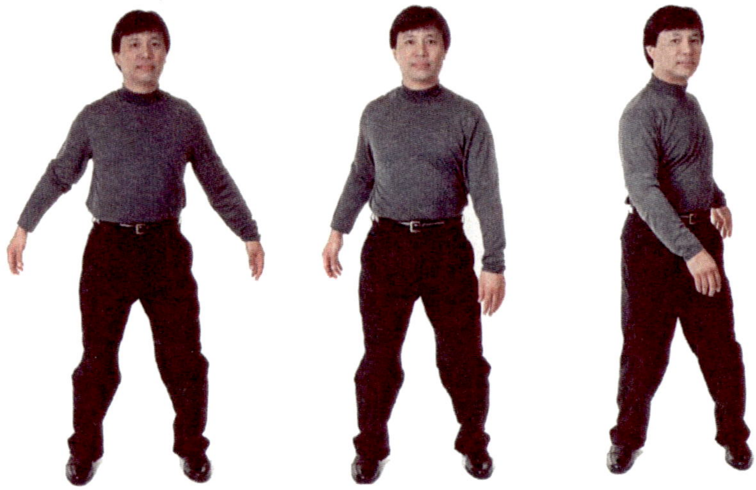

Starting Position

Shake your whole body to loosen up, and build up the rhythm.

Figure 12-2: Eye Therapy

Starting Position

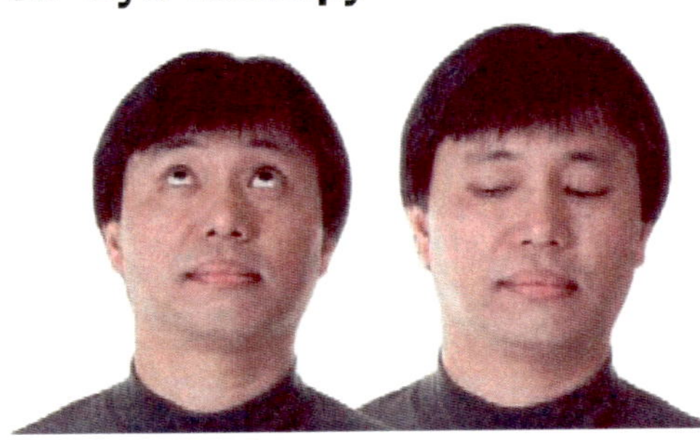

Move your eyes up & down (10 times)

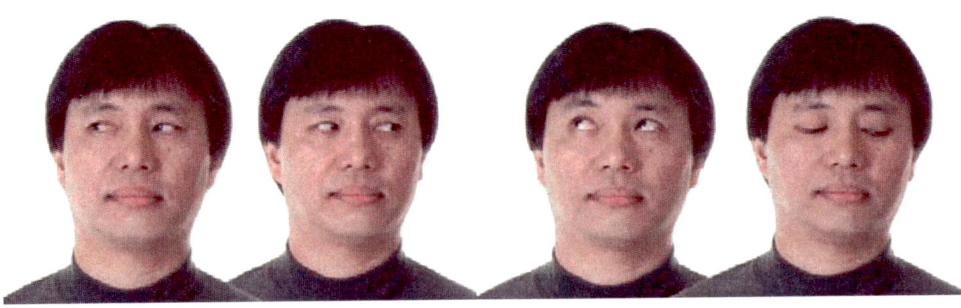

Move your eyes from side to side (10 times)

Move your eyes diagonally (10 times)

Figure 12-2 Eye Therapy

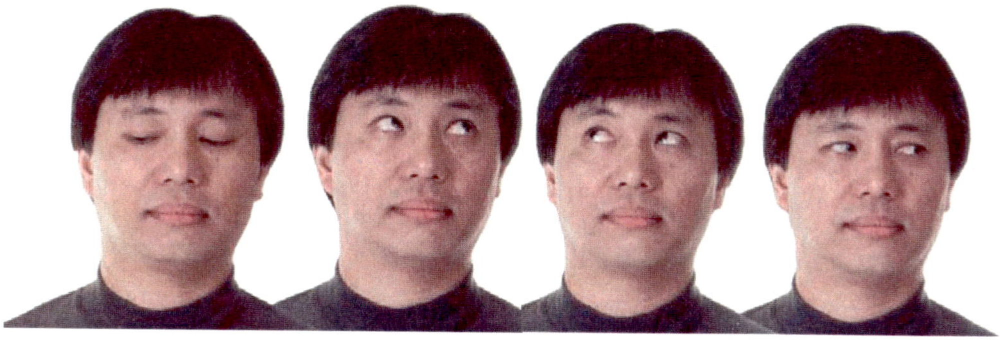

Move your eyes diagonally (10 times) (opposite direction from previous set)

Rotate your eyes counterclockwise 10 times, then rotate them clockwise 10 times

Figure 12-3: Neck Therapy

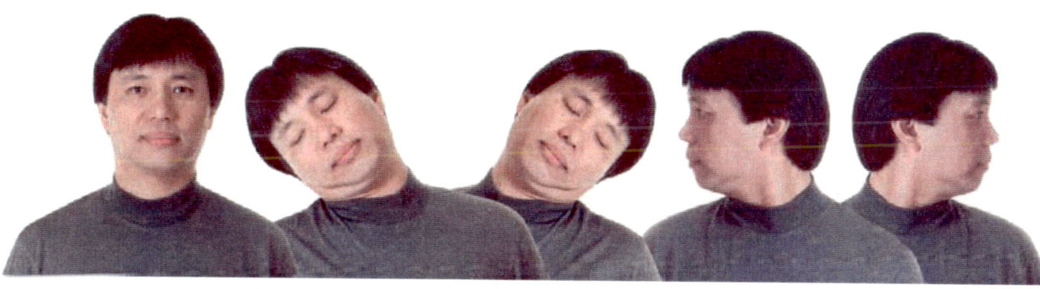

Starting position

Tilt your head from side to side (10 times)

Face to the right first, then to the left (10 times)

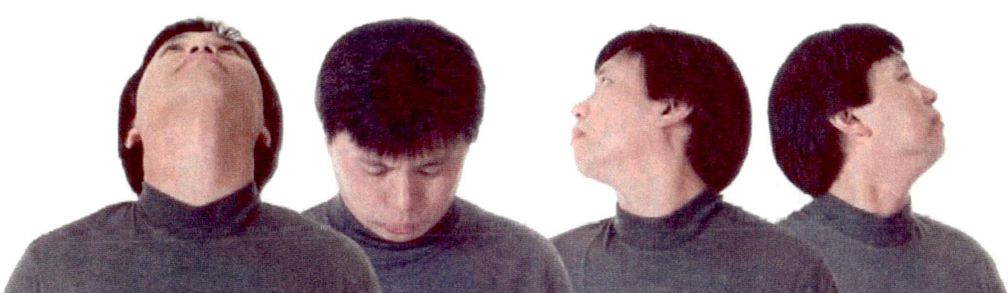

Move your head up and down with full breathing (10 times)

Rotate your head slowly in each direction (3 times)

Figure 12-4: Shoulder & Spine Stress Release

(10 times)

Starting Position

Inhale rapidly as you pull your arms towards the back with your head tilted backwards and your back arched. Then exhale rapidly as you curl yourself forward, chin in and returning your hands back to the front.

Figure 12-5: Arms & Hands T-Exercise

(10 times)

Starting Position

Inhale fully as you move your arms to the side, hold your breath and close-open your hands 3 times.

Rotate your hands around your wrists 3 times, then exhale fully as you return your arms to the front position.

Figure 12-6: Arm Swing

(10 times)

Starting position

Inhale fully as you swing your arms around your shoulders by circling them up towards the back, then exhale fully as you bring them back down to the front.

Figure 12-7: Butterfly Stroke

(10 times)

Starting position

Inhale fully as you circle your arms back and downwards around your shoulders, then exhale fully as you bring them back to the front.

Figure 12-8: Upper Body Twist
(10 times each direction)

Starting position

Inhale fully as you twist to the far right and exhale as you swing back to the left. Do this 10 times, then reverse the direction and breathing for another 10 sets.

Figure 12-9: Hip Rolls
(10 times each direction)

Starting position

Rotate your hips to the right in a circular motion 10 times, then reverse direction 10 times.

Figure 12-10: Internal Organs Exercise
(10 times)

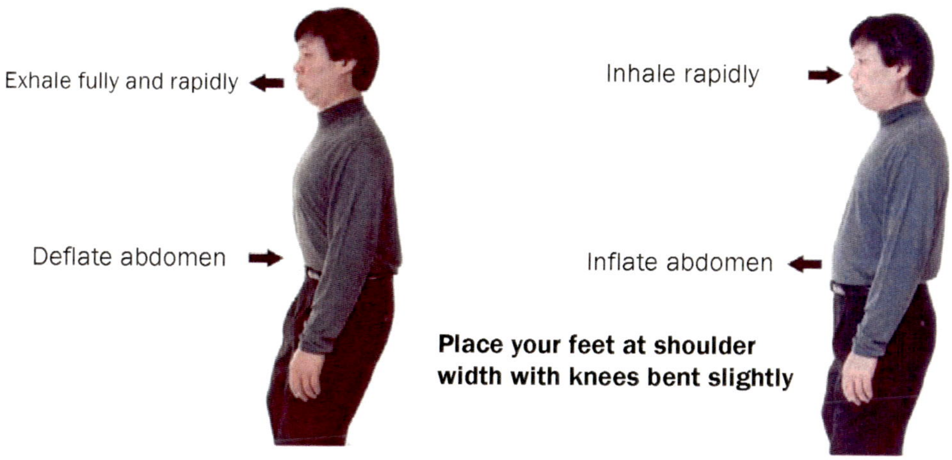

Place your feet at shoulder width with knees bent slightly

Exhale rapidly while compressing your whole abdominal area, then inhale rapidly while expanding the abdomen.

Figure 12-11: Expanding Squats
(20 - 30 times)

Starting position **Downward squatting motion** **Side view of vertical spine while squatting**

Inhale fully in the starting position, then exhale fully as you squat down. Inhale fully as you return to the starting position.

Figure 12-12: Whole Body Stretch

(10 times)

Starting position

Inhale fully as you pull your hands up as high as you can, stretching your whole body on tip-toe. Then exhale fully as you return to starting position.

Figure 12-13: Side-to-Side Body Stretch

(10 times)

Starting position

Inhale fully as you pull your hands over your head. Hold your breath and tilt from side to side. Exhale fully as you return to the starting position.

MDP Body Therapy & Inner Power Development Program Procedure: Part 2

Inner Power Development Technique

This inner breathing technique has a great overall impact on the development of inner powers and internal energy. Certain faculties are particularly enhanced using specific hand positions from internal martial arts and advanced yoga. These steps were designed to activate and revitalize important energy centers of the body:

1. **Navel energy centers** - for internal energy, stamina, good instincts and sexual vitality
2. **Heart energy centers** - for love, compassion, happiness and inner peace
3. **Paired heart and navel centers** - for integrated emotional and instinctive intelligence
4. **Ajna or mid-brow centers** - for improved concentration, mental stamina and abstract mental intelligence
5. **Crown energy centers** - for intuition, divine inspiration and wisdom

Developing and refining the above qualities through the manipulation of the energy centers improves personal success and inner growth. Regular practice gives you more vitality and the gradual development of inner faculties, which will improve your ability to perform better and achieve your life's goals.

The inner power development technique uses hand *mudras*. In Sanskrit, a *mudra* refers to a special hand position and arrangement to seal or flow energies. If you observe the pictures of Saints from different religions, their hand position or *mudra* is the two palms together at the chest with fingers pointing upwards (see Figure 14 on page 71). This is called *anjali mudra* in the East, where it is used by monks and yogis for deep prayer or meditation. Its effect when used is to render the person receptive to the Soul's energy and to facilitate the heart energy center to be connected to the Soul, allowing transformation of consciousness.

Inner Power Development Procedure

After completing the Body Therapy exercises, proceed as follows:

1. **Close your eyes. Do the heart-hand *mudra* or *anjali mudra*** by joining your palms together at the center of your chest, with your fingers pointing upward.
2. **Navel breathing**: Then swing your hands down to the navel with your fingers pointing downward. The sides of your palms should be touching the navel area. This technique causes higher energies to descend to the lower parts of the body. The energy flows wherever your hands are directed. Concentrate on the navel center while breathing slowly and deeply for 1 minute. This will activate and energize the navel energy center.
3. **Heart breathing**: Swing your hands back from the navel to the heart energy center at the mid-chest. Concentrate on the heart energy center at the mid-chest area while breathing slowly for 1 minute. This will activate the heart center.

4. **Heart-navel breathing**: Maintain your left hand at the heart and swing your right hand down to the navel. Maintain this position while concentrating simultaneously on the heart and navel centers and breathe slowly and deeply for 1 minute. This technique will enhance and align your emotional and instinctive intelligence – good for decision-making faculties.
5. **Navel-heart breathing**: Switch your hand positions so that the left hand swings down to the navel and the right hand swings up to the heart. Maintain this position while concentrating simultaneously on the heart and navel centers and breathe slowly and deeply for 1 minute. This will activate, energize and balance the heart and navel centers.
6. **Energy flow**: Swing the left hand back to the heart center, joining it with the right hand, fingers pointing upward. Swing the joined hands down to the navel and back up to the heart repeatedly 18 times. End at the heart center. This technique facilitates the flow of energy down to the lower energy centers and then upward until the lower and higher energy centers are balanced. This technique is very important to circulate and balance the human energy system.
7. **Crown-heart breathing**: Raise your left hand up to the top of the head at the crown center with your fingers pointing upward. Your right hand maintains its position at the center of your chest with fingers pointing upward. Maintain this position while concentrating simultaneously on the crown and heart centers. Breathe slowly and deeply for one minute.

This will activate, energize and align the crown and heart centers. This technique stimulates and integrates the faculties of love, wisdom, intuition and universal consciousness and also improves the connection of the physical, emotional, mental bodies to the Soul.

8. **Inner salutation to loved ones and others**: Bring your left hand down to your heart. Join both hands together at the center of your chest, fingers pointing upward. Bow your head slightly. Feel love, gratitude and respect in your heart. Then for 1 minute silently express these qualities to your loved ones and family members. Do the same for those who have helped your life for another minute, expressing yourself as if this were your last opportunity to do so. Then express good feelings and wishes to your friends, co-workers and associates for 1 minute. Silently express an apology to people who have been hurt by you. Then silently forgive those who have hurt you in the past or are still hurting you. Mentally express your opinions to them without judgment. This technique and process will build a positive program for better relationships. Extend forgiveness for at least 2 minutes.
9. **Inner salutation to the Source of Life**: Keeping your palms together with fingers pointing upward, bring both hands up to the forehead area. The sides of your palms should be touching your forehead. For 2 minutes, silently express your love, respect and gratitude to the Source of Life, the Divine Macrocosm, as if this were your last opportunity to do so. If you have an affinity with Spiritual teachers, Saints or Sages, you may include them in your inner salutation. Enjoy the positive effects of this technique.

10. **Gently bring your hands down to the heart center**, palms together, fingers pointing upward.

11. **Open your eyes and smile**.

12. **Enjoy the pleasant feeling, well-being and balance of aligning yourself.**

The MDP Body Therapy and Inner Renewal Program is recommended at least 3 times a week. Do it before you go to work, after your morning shower to empower yourself for the day. It is also helpful after a stressful day to revitalize yourself before spending time with your family. Try the program for one week to obseve how you feel.

Figure 13: Inner Power Development Procedure

(10 minutes)

Inner Breathing with Hand *Mudras* to Activate and Align the Power Centers

Starting position ***Anjali Mudra***

Navel Breathing (1 minute)

Heart Breathing (1 minute)

Heart-Navel Breathing (1 minute)

Navel-Heart Breathing (1 minute)

Energy Flow

Return to the *Anjali Mudra* and swing your joined hands down to the navel and back up to the heart repeatedly 18 times.

Crown-Heart Breathing

Start from the *Anjali Mudra*. Raise your left hand up to the top of the head. Concentrate on your heart and crown centers while breathing slowly for 1 minute.

Inner Salutation
to Loved Ones & Others

Return to *Anjali Mudra*. Bow your head slightly and silently express love, gratitude and respect to your loved ones, family members and others for 2 minutes.

Inner Salutation
to the Source of Life

Bring both hands up to the forehead area and silently express your love, gratitude and respect to the Source of Life for 2 minutes. Then return to starting position.

Anjali Mudra **with 2 hands joined in front of the chest**
Figure 14

The science of breathing for inner power development

Breath is life. People can abstain from eating food for a week and not drinking for a few days, and they can do without sunlight for a month. But no one can go more than a few minutes without air. Why? Because of the absence of oxygen. Not only that: the vitality and energy that vivify and revitalize the physical and etheric body with all its systems and organs depend on the energy that comes from the air. Air, therefore, is not just a mixture of oxygen and nitrogen and gases, but also includes a special planetary life we earlier referred to as air vitality globules, which are tiny particles of light constantly moving in the atmosphere. By breathing air, we breathe both the essential oxygen and air energy vitality that energizes the physical and etheric bodies.

Air = O_2 + N_2 + CO_2 + toxins + air vitality globules

How the Body Uses Oxygen
A. Metabolism - Oxygen is needed for the proper transformation of food nutrients into energy harnessed by the cells in an internal process. It is the ability to convert food into energy that makes the food we eat nutritious. People can eat the healthiest food in the world but experience poor conversion because of a low level of oxygen. This results in low vitality and malnutrition.
B. Oxygen and air vitality globules properly absorbed through the lungs by full breathing techniques enable the blood to displace the toxins in the body, especially carbon dioxide, and the absorbed air globules are distributed to the chakras and the etheric body. Therefore, continuous revitalization and detoxification occurs automatically.

C. Rapid breathing techniques energize and empower the nervous system through the absorption of ions in the ovula area of the mouth via nerves at the turbinal bone structure in the internal nasal cavity. Therefore, certain rapid breathing techniques can be employed to facilitate this process.

Breathing and body posture

Since breathing techniques are especially effective methods to energize the body rapidly, let's discuss the significance of breathing to health and power. Posture is one of the most important factors impacting the effectiveness of the breathing process.

Proper body posture not only determines air intake capacity, but also one's mental, emotional and spiritual aspects. Observe the sitting habits of different people. How do they generally sit and breathe? These are my observations:

a. **Successful and smart people** generally sit upright most of the time, especially when they are thinking or doing their most important work. They have better breathing habits except when they get stressed; then breathing becomes shallower and erratic.

b. **Depressed patients and pessimists**. Their posture is a concave, curled body when sitting or sleeping. This posture results in a compressed diaphragm, reduced lung capacity and less oxygen intake, resulting in low vitality and reduced power.

c. **Stressed people and patients with anxiety**. These people generally breathe through chest movement and expansion while the abdomen is tucked in, accompanied by a shallow and erratic breathing pattern. Again less air intake means less oxygen and air globules, resulting in poorer health. If this type of person has been exposed to severe stress and anxiety for an extended period, they have the habit and pattern of shallow breathing even while resting or relaxing. Proper, healthy breathing involves abdominal movement and expansion to draw more air deeper into the lower lobes of the lungs, meaning the abdomen expands, not the chest.

d. **Insecure and defensive people**. They usually sit with crossed legs and arms. If you see people in this position, they aren't in a receptive mood, so it's best not to present your ideas and recommendations until they uncross their arms and legs. The closed position does not allow them to absorb vitality from the environment or from the power of ideas during a lecture.

For example, my classmates who didn't excel in school usually sat in this position while listening to the teacher. Do not continue negotiating or trying to give advice to your business clients, colleagues or even to your children when they are standing or sitting with arms and/or legs crossed.

However, this position can be useful on occasion: When you are with very negative and stressed people or in a negatively charged environment, crossing your arms and legs is an immediate technique to shield yourself.

Experiment with this concept:

1. Businesspeople: Test the receptiveness of clients sitting in this position.
2. School teachers: Observe the students who do well and those who under perform. How do they sit?
3. Health care professionals and healers: How do sick patients sit compared to healthier people?
4. Coaches or team leaders: How do high achievers sit compared to pessimists or lower performers?
5. Business leaders: How do your best employees sit compared to under performers? How do your stressed and unhealthy employees breathe? Observe the abdomen and chest.
6. Parents: Observe how your children sit. Who is more active and healthier?

Wrong body posture can create negative physical, emotional and mental health and attitudes, or vice-versa. By positioning the body properly while sitting, standing or sleeping, plus breathing properly with the abdomen instead of the chest, greater health and substantial internal power and energy ensue. The physical exercises and meditations presented in this book always include breathing techniques.

Let's expand on our discussion above. There are five general types of breathing:

Five general types of breathing

First type of breathing: Abdominal breathing

This natural method involves expanding the abdomen during inhalation and deflating the abdomen during exhalation. Observe the natural breathing of a sleeping baby. The belly expands during inhalation and deflates during exhalation.

Figure 15: Proper Abdominal Breathing

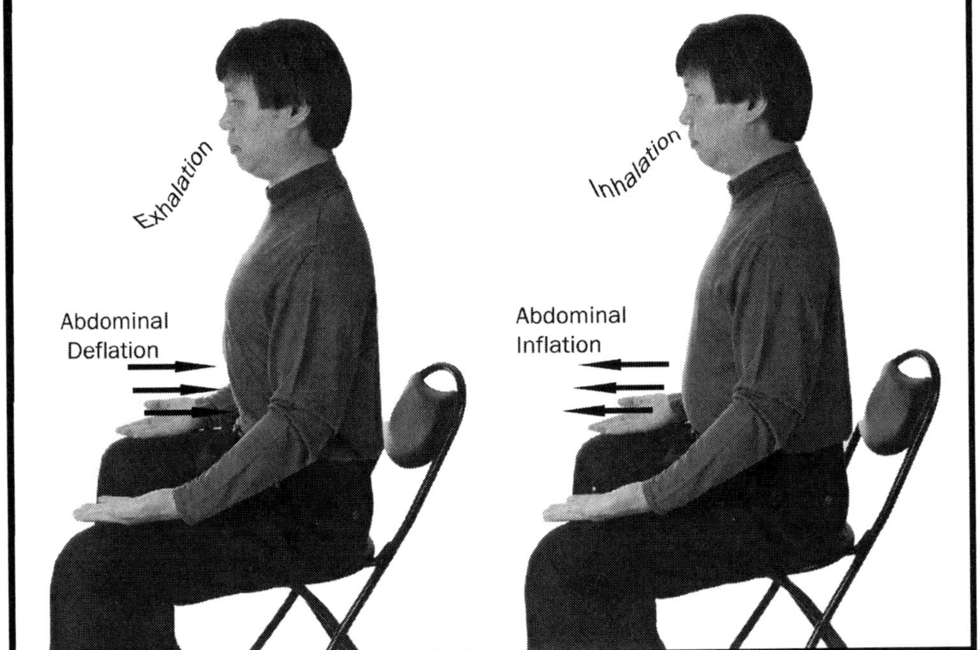

Before we discuss methods to train proper abdominal breathing, Figure 16 demonstrates the sitting positions you will be using for techniques throughout this book.

Figure 16: Positions for Meditation and Breathing Practices

- **Semi-lotus position with spine erect**
- **Seated on a chair with spine erect**

Most adults need to retrain themselves in proper abdominal breathing. Abdominal breathing can be practiced daily in two ways:

Slow Abdominal Breathing Technique (1st Type)

Natural abdominal breathing while externalizing negative feelings, body tension or discomfort guided by your intention. This is a self-guided tool for stress or pain release. It is also a safe, gradual revitalizing technique and fast relaxation method. Don't be fooled by its simplicity; it is a powerful technique.

1. Sit with your spine vertical, or if you are lying down, make sure your body is straight and aligned.

2. Breathe naturally and relax. You may close your eyes.
3. Externalize negative feelings, body tension and discomforts just with your intentions while breathing out.
4. Continue to relax and continue the breathing process and externalization of negativity or discomfort.
5. Do this procedure for 5 minutes, daily or 3 times weekly.

Slow Abdominal Breathing Technique (2nd type)

Slow abdominal breathing with breath retention. This is commonly used in yoga and martial arts. This method will increase your vitality rapidly, but it is not recommended for pregnant women and those with hypertension or heart ailments.

1. Sit with your spine vertical and your feet flat on the floor. You may also sit in a lotus or semi-lotus position.
2. Inhale slowly and deeply using abdominal inflation while mentally counting to 10.
3. Hold your breath for a count of 5.
4. Exhale slowly deflating the abdomen completely while mentally counting to 10.
5. Hold your breath for a count of 5.
6. Repeat the whole breathing pattern for a maximum of 5 minutes.
7. Do this procedure daily or 3 times weekly.

These two types of abdominal breathing techniques have been observed to offer the following benefits:

1. Rapid increase in internal energy and vitality
2. Calming of the mind and emotions
3. Rejuvenation of the body

The two abdominal breathing techniques can be used as a simple meditation to de-stress, relax and detoxify the body, then to revitalize and rejuvenate yourself. You can do the techniques daily before and after work.

Second type of breathing: Chest breathing

As we've discussed, most stressed people breathe into their chest, holding their abdomen rigid. When people have anxiety, anger or high levels of stress, their breathing is shallow. This tends to congest the heart energy center and also reduces air and oxygen intake. You can observe the breathing of your family members, co-workers or friends to see for yourself who is stressed or has emotional problems. I don't recommend continuing with this chest breathing pattern if that's how you breathe because this kind of respiration induces anxiety and emotional overreactions, plus it lowers your lungs' capacity to absorb air. Practice abdominal breathing consciously to reverse this habit or use the internal organs exercise method from the MDP Body Therapy and Inner Power Development Program (see page 56).

Third type of breathing: Shoulder breathing

Your respiratory system and body's own intelligence do shoulder breathing automatically to draw more air in when your system needs more oxygen. After a strenuous physical workout like sprinting or lifting heavy weights, your body demands more oxygen and vitality. Therefore, you tend to experience rapid respiration and deep breathing by expanding your upper trunk and lifting your shoulders. The problem is that if you hold your abdominal area while doing shoulder and chest breathing, you are compressing the diaphragm instead of maximizing your lung capacity, which pushes the lungs upward, causing less air supply.

Fourth type of breathing: Combination breathing

If you are running out of air during strenuous exercises, are so weak and devitalized that you are almost fainting or are low on sexual potency and need revitalization, try the following breathing technique. Here's the procedure combining abdominal, chest and shoulder breathing for maximum capacity:

Figure 17: Combination Breathing

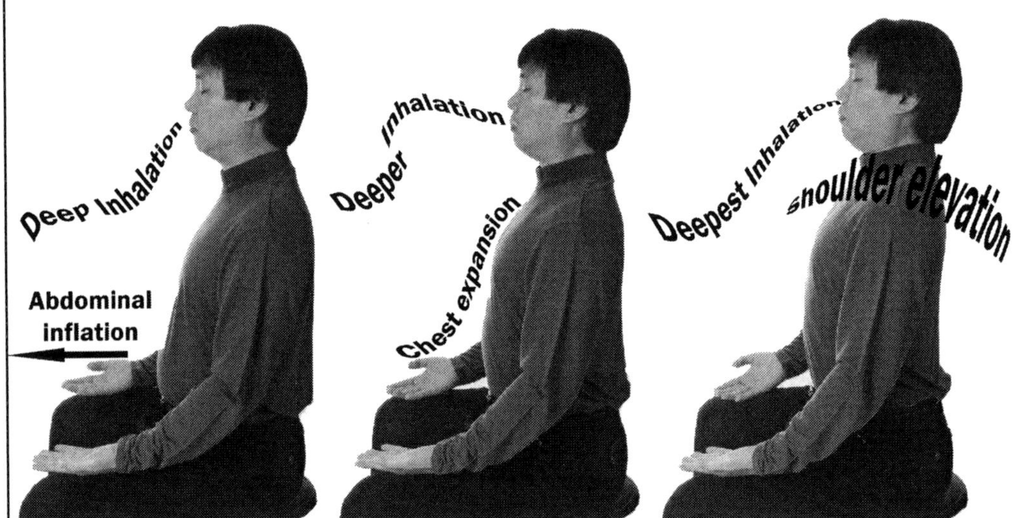

Combination Breathing Technique

1. Sit in a comfortable position with your spine vertical and your feet flat on the floor or in a lotus or semi-lotus position. You may close your eyes.
2. Place your hands with your palms resting in your lap.
3. Inhale slowly and continuously while inflating the abdomen.
4. Continue inhaling slowly, inflating your chest.
5. Continue inhaling until maximum respiration while expanding and lifting your shoulders upwards.
6. Hold your breath steadily for a few seconds (3-4 seconds).
7. Exhale all your air slowly while deflating your abdomen and chest and bring your shoulders back down to their normal position.
8. Hold your breath steadily for a few seconds (3-4 seconds).
9. Repeat this breathing process (steps 3-8) only 5-10 times once a day. Do this 3 times weekly.
10. Precaution: this breathing technique, simple as it may be, creates tremendous internal power and a surge of vitality that is not recommended for pregnant women or people with chest pain, heart ailments, high blood pressure, migraine or headache, cancer or AIDS. If you are not sure, consult your physician before you start this procedure.

Multipurpose breathing technique

MDP Synchronized Breathing Technique™

The MDP Synchronized Breathing Technique was designed for rapid stress release, calming the mind as well as purifying and revitalizing the bodies and chakras. I designed this method after many experiments with the training I received from my early spiritual mentors. Also, while teaching and traveling in India, I encountered several advanced yogis who practice *pranayama yoga*, which literally means yogic breathing science in Sanskrit. Several yogic and healing books also mention two breathing methods: 1) *Kapalabhati pranayama*, which involves rapid inhalation and exhalation with simultaneous up and down head movement 2) *Bastrika pranayama*, which involves rapid compression and expansion of the abdomen during exhalation and inhalation.

Based on my experiments, when I integrated these two breathing styles into one technique, it brought fantastic results: greater detoxifying and revitalizing of the human energy system and rapid calming of the mind and emotions. I have also added chakral breathing to activate the heart and crown chakras. It can be done as follows:

MDP Synchronized Breathing Technique

1. Sit in a comfortable position with your back straight and vertical in a lotus or semi-lotus position or with your feet flat on the floor. Close your eyes before you start.
2. Inhale rapidly through the nose as you tilt your head backwards while simultaneously inflating your abdomen.
3. Immediately exhale quickly through the mouth as you bow your head down while compressing your abdomen.
4. Repeat steps 2 and 3 10 to 15 times.
5. Then just relax with your head vertical and breathe slowly. Concentrate on the top of the head (crown energy center) and center of the chest (heart energy center) simultaneously while breathing slowly. Do this step for 30 seconds.

6. Repeat steps 2 through 5 about 2 to 4 times, daily or 3 times weekly.

This technique will activate and create more energy in the crown and heart energy centers, bringing more spiritual energy to purify and revitalize the bodies. Also, the Soul power can come down to stimulate the faculties of love, compassion, intuition, Divine inspiration, wisdom and universal consciousness in the personality. Integrating the rapid breathing and activation of the two energy centers surpasses many methods to detoxify the human energy system, quiet the mind, calm the emotions and bring alignment of your personality to your Soul. You can achieve great inner power when you are empowered by your Soul and when your mind and emotions are calm and centered. I have used this technique successfully with many executives, leaders and spiritual practitioners for stress management and meditation.

You can use this method for several practical applications:

1. To clear up your emotions and mind before decision-making, negotiating or problem-solving. Mistakes usually occur when you are negatively affected by stress, anxiety, anger, fears and other negative emotions and thoughts.

2. For advanced meditators or beginners, you can use this technique before meditating to help purify and increase the rate of vibration in your bodies. The success of calming the mind and emotions and the depth of your meditative experience depend on the cleanliness and level of vitality in your different bodies. When they are purified and aligned, they become receptive agents for the high incoming vibrations, ideas and transformational energies.

3. The clarity and accuracy of psychic or intuitional perception and interpretation greatly depend on the purity and vitality level of the reader's aura and energy centers. Many of them give wrong interpretations of what they see or perceive psychically due to negative contamination in their energetic, emotional, and mental bodies. Their perception is colored and influenced by their own emotions and thoughts, sometimes in the form of beliefs, inclinations, illusions, emotional preferences and personal experiences. Therefore, if you are a "seer" or psychic reader doing readings, the MDP Synchronized Breathing Technique will not only give more accurate "readings", but it will also make your psychic faculties and intuition easier to use. By activating the crown and heart energy centers, your Divine connection or your "spiritual antenna" is stimulated to a higher degree, thus the Soul or Higher Self can guide your personality better. Divine energy improves the functioning of the mental and emotional faculties.

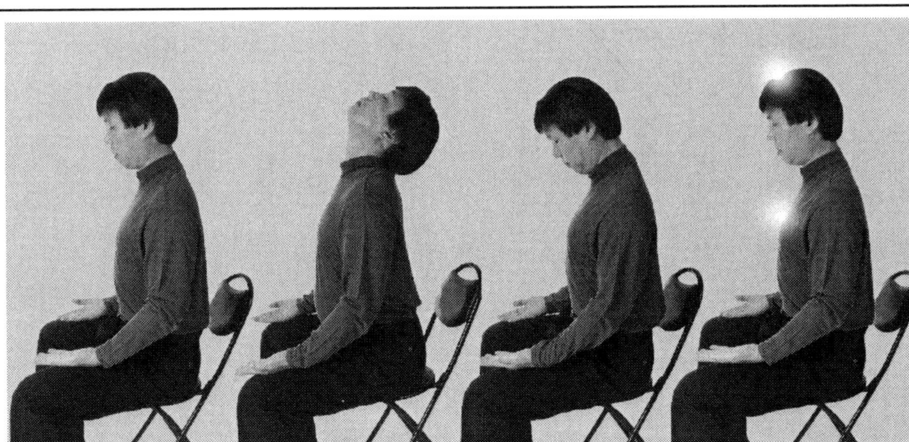

Starting position

Inhale rapidly as you tilt your head backwards while inflating your abdomen, then exhale rapidly as you bow your head down while compressing your abdomen. Do this 15 times. Then concentrate on your heart and crown chakras and breathe

4. Healers, massage therapists, chiropractors, osteopaths, body workers and health care professionals can also benefit from the MDP Synchronized Breathing Technique. The quality of the practitioner's energy – spiritual, mental, emotional and etheric – affects the quality of service offered to their clients through energetic osmosis via the interaction of the practitioner's and client's auras. Our modern times have pushed almost everyone into stressful duties, time constraints, relationship pressures and economic instability. Even health care workers and spiritual servers are stressed and sometimes are not emotionally and mentally stable. If the health practitioner has negative energy and psychic contamination, these can be transferred to clients or patients. I recommend practicing the MDP Synchronized Breathing Technique for a few minutes before meeting with each client, especially for healers and body workers. Medical practitioners might also want to do this technique a few times daily. The clients and patients deserve and will receive a higher quality of energy to recover faster and get healthier.

Just about everyone can use the MDP Synchronized Breathing Technique as a practical tool to maximize performance in work, relationships and spiritual practices. It will also improve your overall health.

> Precaution: The technique is not recommended for pregnant women or those with neck injuries, high blood pressure, migraine or severe heart conditions. If in doubt, consult your medical doctor before you perform this technique.

The power of vitality is what most people need nowadays, whether young or old, healthy or sickly, dynamic or depressed. This section is dedicated not only to unhealthy people so they can recover fast, but it is also designed to meet the needs of healthy individuals to become even healthier and more powerful. In fact, the methods were created to be simple, easy-to-apply and ready-to-use without the complication of needing special gadgets, attire or place to practice. My wish is that you can use them weekly as a way of life.

The techniques to enhance vitality are a simple science once you understand the following:

1. The etheric or vitality body controls and energizes the physical body. If you condition the vitality and health of the etheric body, you rejuvenate the physical body. The energy level of the vitality body is directly proportional to the energy and power of the physical body.

2. The etheric body has chakras that control the health of the organs and systems of the body. By purifying and revitalizing the chakras through exercise and breathing techniques, you empower the condition of the organs and body systems.

3. The etheric chakras have a relationship with certain inner powers developed. Thus by activating them through breathing techniques, you develop certain powers corresponding to the chakral faculty.

4. There are many techniques assembled into menus or methods applied to various requirements. You can choose different techniques for specific applications. But there are at least 3 important techniques recommended to be done weekly as a way of life:

Weekly vitality management schedule

a. MDP Body Therapy and Inner Power Development Program (daily or 3 times weekly)

b. MDP Synchronized Breathing Technique (daily, 3 times weekly or as required to release stress or calm the mind)

c. Slow abdominal breathing or combination breathing (2-3 times weekly)

When you master these techniques, you substantially master your vitality power.

Vitality mastery and management are not just about adding energy. It is not just building physical muscles or stamina. Sometimes, you need to subtract negative energy first, like stress and negative feelings and thoughts. It's about subtracting negativity in your system in order for you to add positive energy and vitality. You need to include a regimen to rapidly release stress and tension in just a few minutes and to bring back vitality and stamina fast. Isn't this what you arc looking for: fast, simple and effective? You've got it!

Section B
Emotional Power & Intelligence: Principles & Techniques

I. **Ramifications of emotional powers**.................................... page 90

A. Powers of passion, leading to courage, determination & commitment....................................... page 91

B. Power of the heart: Love & compassion........................ page 93
   1. Technique for energetic self-defense.............. page 93
   2. Emotional powers of great world leaders.. page 95

C. Techniques to build emotional power & intelligence... page 97
   1. Inner Purification Technique: Part 1....................... page 97
   2. Inner Purification Technique: Part 2....................... page 103
   3. Alternative method... page 104

D. Power of emotions.. page 104

E. Power of emotional magnetism and presence............ page 106
   1. Magnetic Presence Technique................................ page 108
   2. Magnetic Presence Technique precautions.. page 111

Section B
Emotional Power & Intelligence: Principles & Techniques

There are different grades and types of emotional power and intelligence, ranging from strong passion that pushes people to achieve their goals and passion that characterizes sentimental relationships to the fearless devotion that causes people to be daring or courageous, sometimes to the point of extreme fanaticism. There is also the emotional aspiration of great leaders, which brings strong adherence and commitment to higher ideals and goals. At extreme levels, this can lead to a martyr complex. Emotional intelligence expressed through balanced love and devotion to great ideals and goals, accompanied by compassion and understanding, is an inner power that can lead to great achievements. This emotional power results in a quality and ability to be liked by people and, if you are a leader, to be cherished by your followers or team. This is sometimes called emotional magnetism.

What makes a great influential leader powerful is not just their expertise or their power derived through position or office. One of the most important qualities they have is the ability to be liked by their subordinates, peers, followers or the public. What causes this? Let's look at a celebrity as an example. I have been studying why some television personalities excel and have high ratings year after year. We can learn through their style of using not only talent but also inner powers. Often, they're not even aware of these inner powers.

Oprah Winfrey, a US talk show host, actress and producer, is a good example. Aside from her genius as a creative communicator with her audience, she is also able to build rapport by exuding an aura of attention and compassion to the needs of her viewers. This is because her heart power center is developed. She uses a heart-to-heart talk style, transparent and direct. Her talk show format is similar to Phil Donahue's, who I think helped pioneer the American talk show format of talking to the audience, but what's different about Oprah is her emotional and personal quality to connect to the audience and viewers. She uses the power of loving emotions and the ability to be liked by people through the faculties of the heart.

When a clairvoyant or psychic looks at Oprah's aura and energy centers, they can perceive that her heart energy center is active and open. She has a big heart, and this magnetic link with and radiation to her audience is natural and very attractive. I have read that she also shows this through her ongoing philanthropic projects and work supporting charity and education. Therefore, her practices entitle her according to universal principles like the Law of Cause and Effect to achieve great things in life.

Among other international names, Mother Teresa and Princess Diana are also good examples of people using the power of emotional intelligence as a natural tool to be liked. Mother Teresa and Princess Diana were loved by the media and the international public. Mother Teresa's compassionate heart made her a candidate for canonization as a Saint. Her powers of servant leadership, inner values and compassion left a legacy for her followers.

Ramifications of emotional power and intelligence

As with many aspects of life, the power of emotions can lead to positive or negative results. Different levels of emotional expression result in many different types of power:

Ramifications of Emotional Power & Intelligence
Passion leading to courage, determination & commitment
Power of the heart: Love & compassion
Emotional magnetism & presence

First Ramification of Emotional Power

Power of passion, leading to courage, determination & commitment

Most of humanity today has already developed and acquired this type of emotional substance, resulting in the tremendous successes of modern history. On the other hand, our lives in many cities are full of stress created by the desire and determination to be successful, to be Number 1, to be great, to be rich, to be famous and sometimes just to survive modern life. But this suffering isn't all bad; it's a way to transition to a higher goal and purpose, enhancing the will-power of humanity to handle greater responsibilities in the future.

> Desire =
> Lots of emotional urges or impulses,
> accompanied by slight mental intention

When I speak of passion, it is connected to desire, which can be good or bad, leading to either success or failure.

Passion is a combination of emotional and mental energy and force, but based more on feelings or emotional substance. When people have great passion, they possess a dogged determination and courage to achieve their goals because of the desire for success. Many heroes and super-achievers in world history exemplify this power. Many people throughout history have sacrificed their family, career, health and future and have even died for their chosen cause or ideals – money, business interests, love affairs or love for a country. People driven by passion to acquire power and position can sacrifice not only their own lives, but the lives of others as well.

On a more individual level, the lower passion of love is colored with sexuality, jealousy or possessiveness that binds people to be too dependent on personal relationships and eventually leads to frustrations. Many people hurt their partners and cause destruction in families or groups because of their lower type of emotional desire.

Not all objects of passion are noble or great causes. The first underlying test of whether a subject of passion is good is to examine the underlying intentions or motives behind the idea or desires. If the goals are selfish, greedy and self-gratifying at the expense of others, then the passion leads to negative and self-defeating outcomes. Let me distinguish here between selfishness and self-interest. There is nothing wrong with self-interest. In fact, self-interest is an ingredient for people to achieve great results and success. It is through self-interest that people are primarily motivated to perform and work hard. Selfishness is something else. It brings suffering to oneself and to the people affected by their deals and transactions. It is not a win-win situation.

Lower passion is colored by the emotions of fear, insecurity, anger, hatred, jealousy, envy, self-centeredness, self-importance and selfish intentions. These all lead to a miserable and stressful life. So the downfall of emotional powers derives from the abovementioned negative emotions, inner qualities and habits. When a clairvoyant or psychic looks at the aura of those afflicted with these emotional diseases, they see a dark grayish-red and really engrained negativity in the emotional energy bodies of some people. Their aura looks muddy red, dense and energetically sticky with many negative vibrations in the aura and negative entities in their energy centers. If you are energetically sensitive and receptive, you can sense people who belong to this group. You immediately feel different symptoms like nausea or a nervous tension in your stomach and solar plexus area. You also get irritable and devitalized, as if you are getting congested and negatively charged with certain feelings. If you happen to sit beside this type of person, you can do a few things to defend yourself energetically:

Technique for energetic self-defense

1. Cross your arms and legs and focus on heart and crown chakras while breathing slowly. This will disconnect your aura from the negative people or environment and make you feel detached.
2. Divert your attention to something else like a good visual or painting around you, or recall a happy experience in your life. This will withdraw the energetic connection.
3. Mentally wish these people healing and transformation for their lives.
4. If you pray, invoke that they receive spiritual blessings.
5. Do not entertain any anger, internal dislikes for them or fear of them. If you do, it will connect your aura, and you will become more affected. Practice detachment.

Second Ramification of Emotional Power and Intelligence
Power of the heart leads to love, compassion & aspiration

This higher emotional power is a transformation of the love of things, goals and personalities into a higher intention and motive to use love for greater universal service and better relationships. It involves not only the heart, but also the crown energy center. If a person's desires and prayers are limited to personality or family needs, the development of inner power takes place in the heart and solar plexus energy centers. The solar plexus energy center is where personality and self-interests are stimulated, involving physical appetites and the virtue of courage.

Once the person starts evolving and the heart chakra is more active, intentions and prayers become a wish or invocations for groups, the community good, organizations, cities and countries. It culminates in the selfless sacrifice and service for the good of humanity and the Earth. Thus a person who has mastered this inner power has integrated the faculties of universal consciousness, wisdom, intuition and inspiration.

This can be seen energetically as crown energy center development. Now we understand why the symbol of universal power and enlightenment is the open crown of the Christ, the headdress of the Buddha and the Pharaohs, the crowns of ancient wise kings and queens, the mitre of the Pope and the many-feathered headdress of Native American chiefs. The inner powers of love, compassion and aspiration are the higher counterparts and transmuted power of passion. Normal people begin with passion.

As people evolve into more advanced Souls, they guide their lives with aspiration, compassion and all-inclusiveness. At this high level of heart development, great human beings always leave humanity and the world a legacy of monumental projects and service. The inner power of compassion and aspiration leading to inspiration is the quality of Saints and Sages that lubricates their world service.

What are the common qualities of the greatest world leaders? Let's look at the lives of the Christ, The Buddha, founders of religions, as well as Mahatma Gandhi, Mother Teresa and some modern science and business leaders.

Emotional Powers of Great World Leaders
Universal love (love that exceeds love for self, family and friends)
Compassion for all that lives (not only for humanity but for all kingdoms of nature)
Inner Peace leading to serenity (undisturbed continuous Peace)
Continuous joy leading to bliss (a deep experience attributed to the fulfillment of one's higher destiny and life's purpose)
Altruism (a natural gift of the heart to give and share selflessly)
Benevolence (a selfless act of love and kindness)
Wisdom of the heart or conscience (the power to feel what is right and wrong through your heart) This faculty and inner power is usually called conscience.
Unwavering determination (starts with emotional commitment partially colored by passion and desire, then transmutes into spiritual aspiration and devotion characteristic of many great super-heroes, successful leaders and Saints) This quality leads them to be unstoppable and inspires their great sacrifices.

Some of these qualities contribute to what we normally call "emotional power and intelligence". It is a faculty of the heart and crown centers. How do we know it really exists? Almost everyone can feel a pleasant, warm opening feeling in the heart when one does good things for others. But when a person hurts or does harm to someone, either through negative thoughts, feelings or actions, one can feel a shrinking, heavy, unpleasant feeling in the heart or chest area.

When people develop the heart energy center through constant kindness, generosity and goodness to others, the sensitivity of the heart or conscience is enhanced. That is why some pictures of Saints from different religions show a common bright light emanating from the center of the chest. Also many pictures and sculptures of Jesus the Christ, the Buddha, some Hindu Deities and Divine personifications of Holy Mothers show their left hand and fingers touching their heart center while the other hand blesses.

When the heart center is very developed and open, the person is always loving and very sensitive to the needs of others, resulting in a natural expression of compassion. When the compassionate act or expression is extended to a greater influence and global cause, the crown center starts to develop further, resulting in universal love and global service. The ramifications of this inner development include the power of the Divine Self automatically guiding the person with the faculties and virtues of wisdom, benevolence, altruism, intuition, bliss and serenity, which are powers and qualities of Saints from all different religions and of Sages. If you see the auras of Saints and many Sages clairvoyantly or psychically, they have one thing in common – a lot of golden-pinkish and violetish colors in their auras.

In the future, many people including incoming advanced children and Souls will display their psychic faculties and publicly declare the existence of these subjective principles and inner realities. They will easily determine who is who on the planet without pressure or opinionated alterations by the media or publicists. Actually, the role of the media and publicists will change in the future to include seeing and evaluating truth guided by inner reality and psychic faculties. We'll discuss later in Section D how to develop psychic powers like clairvoyance in order to possibly develop psychic abilities and see subjective realities for yourself.

Techniques to develop emotional power and intelligence

Inner Emotional Purification Technique

Having discussed the different energy bodies, including the emotional energy body, the heart and crown centers and their effects on the power of emotions, let's now study and apply the concept by learning how to develop emotional power and intelligence.

How can you gain and develop the power of emotions and emotional intelligence safely and properly?

First, release negative feelings and purify the emotional body and its energy centers through the following inner breathing technique:

Inner Emotional Purification Technique: Part 1

1. Imagine yourself in front of a beautiful ocean and expel all negative feelings and emotions by breathing them out. Just have the intention that the ocean is absorbing them. The power of water is demonstrated by:
   a. The Holy Water of the Christian Church, which is consecrated salt and water.
   b. The Dead Sea with its proven healing properties.
   c. The healing effects of saline solutions used, for instance, to cleanse contact lenses.
   d. Many household products with salt ingredients for detoxification.

e. Pranic Healing and some Indian healing methods, which throw negative energies into a basin of salt water. The ocean is a big salt-water solution and has a property of transforming negative substances and energies, even through visualization.

2. Do deep and slow breathing for about 5 minutes, relax and let go. Breathe out the following negative feelings and emotions:
   a. Stress and emotional tension to improve performance
   b. Fear, worries and phobias. These reduce your vitality and weaken your power. Removing them always enhances your virtues and growth.
   c. Irritability, anger or hatred. These deactivate and inhibit your emotional and mental powers and negatively influence your relationships. These negative attitudes drain you.
   d. Jealousy, envy or lack of confidence. These corrupt relationships and slow down your performance. Releasing them allows you to enjoy inner peace and self esteem and builds trust in relationships.
   e. Compulsive behaviors and obsessions. These destroy your life and health, especially if they result in vices. By releasing them, your good health will enable you to enjoy your future and will give you the ability to regulate your life for deeper fulfillment and inner power.
   f. Selfishness and greed. Releasing them gives you the power to continuously materialize and enjoy abundance and joy in life.

You can externalize many minor emotional negative attitudes by breathing them out, but people generally suffer from those above. All negative attitudes and behaviors are just energies, and their derivative entities range from slightly dirty grayish-reddish energy substances to very dark vibrational forms and shapes.

The shape, form and color of fears, for example, are negatively-charged dark grayish energy. The substance of fearlessness is the antidote to fear and is proportional to spiritual attainment. Greed is a dark grayish-reddish energy form normally found in a person's solar plexus, ajna, base-of-spine chakras and parts of the aura. All negative behaviors, including compulsive behaviors, are caused by these energy entities, which can be seen by clairvoyants or trained psychics.

If you see the entity of love clairvoyantly, the color and shape is pinkish-light reddish. The more advanced the subject, the brighter the color. The entities of Peace are generally pinkish-bluish-white and bright in color.

The entities of prosperity also have a form and color: reddish, slightly yellowish and golden. The problem with prosperity today is that the energy and entities of prosperity have been contaminated with greed and selfishness and colored with many bad business motives. It is good that we have an increasing number of generous groups and philanthropists who are neutralizing this setback.

Once the hearts are more active and the crown centers more open in the majority of humanity, the nature and the expression of governments, business, education and health of humanity will shift to unbelievable heights and outer qualities leading to a highly cultured civilization.

Substance of the Virtue of Fearlessness

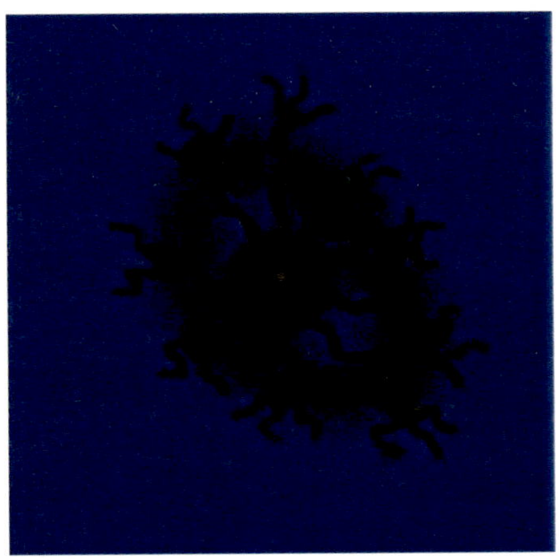

Fear entity

Amplified Portrayal of the Substance of the Virtue of Fearlessness and Fear Entity as Seen by a Clairvoyant

Figure 19

Substance of the Virtue of Love from the Heart Energy Center

Substance of Lower Passion

Amplified Portrayal of the Substance of the Virtue of Love and the Substance of Lower Passion
Figure 20

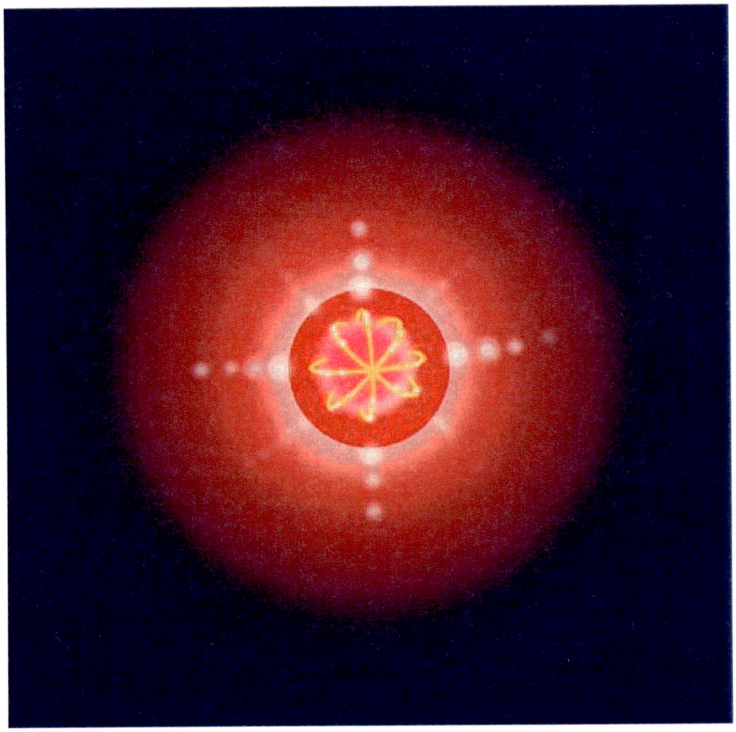

More Advanced Substance of Prosperity

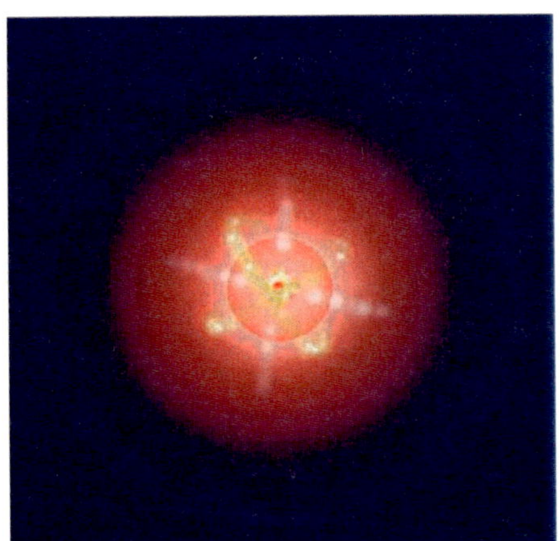

Normal Substance of Prosperity

Amplified Portrayal of Substances of Prosperity
Figure 21

Inner Emotional Purification Technique: Part 2

Substitution, addition and enhancement of positive emotions attitude and virtues

Removing or subtracting the negative emotional attitudes and emotional behaviors by externalizing negative energies that trigger them is not enough. To avoid relapse, we need to substitute positive emotional energies, substances and programs into the emotional body and energy centers. There are two methods to do this:

Inner Emotional Purification Technique: Part 2
To activate and stimulate a positive emotional attitude

(5 minutes, 3 times weekly)

1. Activate the heart and crown energy centers by concentrating on both of them simultaneously while breathing slowly and deeply 7-10 times or about 2–3 minutes.
2. After that, continue the inner breathing and do the following affirmation 3-7 times:
   I am a peaceful, loving and harmonious person. I am a being of love. Love I am.
3. Visualize the positive emotional attitudes or virtues of loving relationships, kindness, compassion, generosity, loving understanding, harmony, and peace as part of your daily life.

Alternative method: Inner purification and activation/stimulation of a positive emotional attitude

(5-10 minutes, 3 times weekly)

Use the MDP Synchronized Breathing Technique (see page 81) and add techniques 2 and 3 above (5-10 minutes). This is a more complete technique that yields fast, wonderful results. I encourage you to experience it immediately before going on to the next topic.

Let's summarize the principles and techniques involved in the development of the powers of emotions:

Power of emotions

First emotional power: The power of passion

The power of passion can empower you with constant success and achievements if it is guided by positive intentions and used for good goals only. The powers of emotional determination, commitment and courage are the main by-products of passion. You can achieve many great accomplishments in life if the power of passion is developed properly based on win-win goals for you and others. When the force of good passion is directed to great causes, you can become a legend.

Second emotional power: The power of the heart

In less developed people, the power of passion is solar plexus- and sex-oriented, especially at the emotional level. In more advanced human beings, passion is a combination of the power of the heart and solar plexus chakras at the emotional and mental levels. This is one of the greatest gifts and attitudes for a human being to acquire, develop and implement as naturally as possible.

Its derivative virtues are: love, aspiration, compassion, energetic magnetism, peace, harmony, joy and a great conscience to perceive truth. Therefore, these inner powers of the heart are extremely important to all roles of your life – in the home and family, in your work and career, in your spiritual duty and life, in your environmental and community contributions and in your health.

Some people's temperaments are not as naturally gifted with emotional powers and intelligence as others, like the mystics. More mental people can develop these emotional powers through the techniques summarized as follows:

1. Inner emotional purification breathing (expelling and removing negative emotional tendencies, feelings and substances)
2. Activating the heart energy center to stimulate the love nature
3. Substituting and adding positive emotional attitudes and virtues through affirmations and visualizations

I suggest doing all of the above techniques at least 3 times a week and whenever you need more positive attitude and emotional power.

The quality of thoughts and emotions that you radiate is what you'll receive back.
Passion begets passion. Anger begets anger.
Divine Love begets unconditional love.

The inner power of the heart is best expressed by a quotation from Blaise Pascal:

> The heart has a reason that reason knows nothing of.
> - Blaise Pascal, French mathematician & philosopher

Third Ramification of Emotional Power and Intelligence

Power of emotional magnetism and presence

Have you ever heard the expressions, "You are very magnetic" or "You're like a magnet"? We've all heard of people with magnetic sex appeal. This magnetic appeal is created by a charismatic aura. Emotional magnetic presence can be classified into the following types:

1. Sex appeal
2. Charismatic emotional magnetism or presence
3. Combination of both sexual and charismatic magnetism, which results in the ability to draw people to you not just with your verbal communication skills, good looks or good public relations. In fact, some charismatic people don't have to open their mouths at all. People tend to stay around them because they feel good in their presence. The real ability to be effective in business transactions and close deals depends on how people feel about you. The product, service and legal contract you're offering might appear good on paper, but how a prospective customer perceives you and feels about you emotionally largely determines the final outcome. The same goes for recruitment and hiring of new employees, especially if you are hiring your assistant or important workers in your organization. It's how you feel about the new hire that matters. Therefore, maintaining a pleasant-feeling, radiating, positive aura and emotional charisma in daily life is very important to relationships, business and careers.

Most business and public relations training has always been based on pleasing the customer and serving the client. The ability to project a good feeling and sincerity in our service is determined by the real inner feelings of the heart center and the vibration of the emotional energy body.

When your heart is developed, you naturally exude pleasant feelings and perform great service that builds server-customer-client relationships.

> You cannot give what you do not have inside you. Love, compassion and forgiveness can only be given when the heart is filled with these positive qualities.

People who are not as advanced have smaller heart centers, so it is hard for them to forgive and forget the pains of life and negative experiences. They do not have as much heart or loving substance to draw upon.

People with small heart centers also get stressed more easily because perception is lower and more pessimistic when the solar plexus chakra is big and the heart is not flowing. People who are considered "big-hearted" are more patient and understanding, so they have greater stress tolerance and better performance.

Improving emotional magnetism

There are several steps people can use to improve emotional magnetism. Emotional magnetism may be developed by enhancing sexual power for attractiveness, anti-aging and energetic creativity. A relationship exists between a person's age and available sexual power. With a technique below, you can maintain sexual vitality even as you grow older.

Attractiveness, rejuvenation and energetic creativity are directly proportional to the activity and level of sexual vitality in the sex energy center.

The degree of awakening of the sacred fires called the *kundalini* (see glossary) determines the level of vitality in the sex energy center and vice-versa. Sexual energy flows from the chakra at the pubic area, so the energy of the perineum and navel centers also affects sexual vitality. I encourage you to activate and vitalize the navel center regularly by breathing slowly while focusing on that area because this will help you maintain sexual stamina and internal power. There are different methods of activating sexual vitality and magnetism, not only for the sake of being physically and sexually attractive, but also to rejuvenate the human energy system for anti-aging. Sexual energy can also be used as an energetic cosmetic for the skin. And maintaining sex energy is important because it enhances creative power and spiritual development since the procreative vitality of the sexual center is a fertilizer for the creativity of the mind.

Magnetic presence technique

(10-15 minutes, once a week)

1. Doing physical exercises before the sexual vitalization technique is preferable to assure quality results. The MDP Body Therapy exercises (see page 58) or a physical workout that at least includes body twists, hip rolls, squats and the body shake are important. Do the exercises for at least 10 minutes. The exercise will help release blockages in the physical and energy body.

2. Perform the MDP Synchronized Breathing Technique for 5 minutes. This will purify the energy bodies and energy centers to assure smooth, unobstructed flow of internal energy and sexual vitality.
3. Focus on the sex energy center by concentrating on the pubic area. Breathe slowly and deeply. Try to hold your breath for a few seconds between inhalation and exhalation. This technique will detoxify and revitalize the sex energy center and rejuvenate the reproductive system. You may visualize the reproductive organs healthy and bright. For women, visualize the ovaries, uterus, fallopian tubes, cervix and breasts healthy and bright. For men, visualize the testicles and prostate gland healthier and bright. People whose organs are removed should do the slow deep breathing focusing on the sex energy center, only without the visualization on the organs. Do this step for at least 5 minutes. For older people, you can do this breathing technique focusing on the sex energy center for 7 minutes.
4. People who have emotional and mental inhibitions or perceptions that sex is dirty and unwholesome need to expel these obstructive ideas and beliefs by breathing them to the imaginary ocean – just by exhaling and with the intention of expelling. These inhibitive beliefs can create a blockage in the sexual center. People who have no negative beliefs or phobias about sexuality, just exhale out stress and emotional discomforts to purify the aura. This can take at least 3 minutes.

5. Now you can empower your aura with sexual energy to enhance emotional magnetism, empower your creativity and rejuvenate your body.

    a. Concentrate simultaneously on the sex, heart and crown energy centers, and breathe slowly 10-15 times.

    b. Position your hands in front of you with the right palm above the left hand. Both palms should be facing down and about 8 inches apart.

    c. Do hand rolls to circulate energy. (See hand rolls in Figure 25, page 214). Exhale all your air first, then inhale very slowly as deeply as you can while rolling your hands forward over each other without physical contact. As you repeat this several times, maintain your concentration on the sex, heart and crown energy centers.

    d. Do step c for 7-10 complete breathing cycles only. Doing the hand rotations and breathing more than the recommended period may result in over-stimulation, over-heating and energy congestion.

Yogis in Tibet and the Indian Himalayas use a similar technique to stay warm in the snow. An advanced version of the technique can melt the snow with the body. I've tried this advanced method several times successfully in my climbs in the Himalayas. This might sound simple, but the technique produces tremendous energy and partially awakens the sacred fires of the body called the *kundalini*. Have you heard the saying, "You are playing with fire"? It means too much heat (from sexual energy and sacred fires) can burn the human energy system with negative side effects.

Magnetic presence technique precautions

This is not only an art but also a science, and healers and yogis have observed negative effects for people with certain conditions or habits:

1. High blood pressure and stroke
2. Heart ailments
3. Tumors, cysts or cancer
4. AIDS
5. Emotional and mental imbalances
6. Sexual diseases, including prostate problems
7. Pregnancy
8. Menstruation
9. Psoriasis, skin rashes or severe skin disorders
10. Smokers, alcoholics and recreational drug users
11. Heavy red meat eaters

Since this technique creates and circulates so much energy, there are other precautions to be observed:

1. Do not bathe or take a shower within 4 hours after doing the technique.
2. Do not eat full meals 3 hours before doing the procedure.
3. Do not drink alcohol 3 hours before the procedure.
4. No sexual release before the procedure to maximize energy results
5. If you are in doubt about whether you can practice this technique or not, consult your physician before starting. If you experience anything unusual that gives you discomfort, stop using the technique.

There are several benefits and rewards of practicing this technique:

1. Increased emotional magnetism. People will feel great around you and will be energized by your positive radiation.

2. The circulated sexual vitality, loving energy and spiritual vibration will rejuvenate the skin and body tissues and will boost vitality of the human energy system for anti-aging.

3. It is a healing and decongesting technique for celibates and people who tend to be sexually congested or insatiable. Energy congestion or stale, uncirculated sexual energy has been observed by healers and clairvoyants to increase the risk of excess tissue growth, cysts and tumors in the reproductive system.

4. The circulated sexual energies will act as a fertilizer to empower creativity. People who have more active sexual centers tend to be more creative. When sex vitality is blocked or weak, creativity declines.

5. It also facilitates the development and activation of the heart and crown energy centers for faster inner development.

6. It can be used as an enhancement of sexual and emotional connection to partners before lovemaking. I am very optimistic that if this technique can be used by teenagers or people who are very sexually active, it will help release the sexual tension and emotional anxiety that cause physical and psychological discomfort.

   For older people, it can help revitalize and invigorate your sexual life and emotional relationships. It's always very helpful to have outer beauty or physical good looks. Of course, physical appearance is the first thing people see in you, but to maintain long-term relationships or rapport with people, it is the inner beauty that counts.

As we grow older, we cannot compete easily with younger people in terms of physical looks. But your inner beauty will maintain your "looking good-feeling good" presence, which will reduce insecurity and fears of growing old and of not being as attractive to your partners or to others. There is not only a "good looking person", but also a "looking good" personality. People with a "looking good" personality possess great vitality and loving, emotional and spiritual pleasantness in their aura. People like to be around people who feel good and radiate warmth and revitalizing power.

Magnetic Presence Technique practice schedule

You can practice the sexual and emotional magnetism technique once a week for adults, twice a week for younger people and once every 10 days for older people. Women achieve best results during their ovulation period when sexual energy is at its maximum level.

More advanced techniques exist to help you improve sexual, emotional, mental and spiritual connectedness and union with your partner or to enhance your inner spiritual experience, but they cannot be published in this book.

I hope that this section on emotional power and intelligence, as you practice its techniques and recommendations, brings you an emotional well-being and the power of love that improves and sustains your ability to be liked by others, allowing your social influence and leadership to grow even further. The culturing of the emotional body and its inner powers has been designed with your needs in mind:

- Minimal or no stress in your daily performance
- Enhanced personal and group relationships
- Better rapport with people
- Magnetic appeal that greatly enhances your external and internal effectiveness and inner beauty
- A wholesome presence, triggered by a pleasant emotional aura
- Right power of passion, courage and determination to help you manifest your life's goals
- Power of the heart as your discriminating conscience for truth
- A life full of love, harmony and peace
- Power of compassion to guide your inner development

The methods were created with simplicity and a straightforward approach in order to suit very busy individuals who have only 5-10 minutes a day to practice.

> For weekly emotional maintenance I recommend the 5-10 minutes of Inner Emotional Purification (page 97) or the MDP Synchronized Breathing Technique (page 81) daily or 3 times weekly.

These two techniques will reduce stress and difficulties by detoxifying the emotional and mental pollutants that are absorbed from other people and the environment. It is like taking an emotional bath to refresh your feelings. Life is good. You should enjoy it more. Continue the practice and keep nurturing the power of your heart.

Section C
Mental Power:
Principles & Techniques

I. Power to physicalize.. page 119

A. Experiment 1: Mind-thoughts relationship to the physical body and its chemistry................................page 120
B. Experiment 2: Mind relationship to emotions & vitality..... page 121
C. Experiment 3: Power of materialization using the mind.... page 122
D. How to program only positive end results........................... page 125
E. Procedures for programming & materializing goals........... page 125

II. Ramifications & elements of mental power........................... page 130

A. Elements of mental power...page 130
1. Common sense... page 130
a. Guidelines for the proper use of common sense..... page 131
2. High grade of common sense... page 134
3. Concrete mind...page 136
b. How the mind works..page 138
- Memorizes & stores information......................... page 138
- Organizing, problem-solving &
- decision-making aptitudes & powers.................. page 140
- Method for problem-solving & decision-making..page 140
- Power of research & investigation...................... page 142
- Mental power to concretize ideas....................... page 143
4. Abstract mental power.. page 147
a. Method to train the abstract mind.............................page 148
b. Power of innovation & creativity.................................page 154
c. Power of imagination as a faculty of the abstract mind...page 157
5. Practical techniques for mental power & its applications... page 164
6. The mental power of focus & awareness....................... page 172

Section C
Mental Power:
Principles & Techniques

There are two aspects of the mind: the concrete and the abstract. The concrete mind deals with the sequential, logical and normal reasoning and concrete thinking faculties. The abstract mind is responsible for the more subjective, creative, philosophical and principle-based thinking faculties and processes. When we speak of the mind, it is the human mechanism or energy software used to think. It is your mental energy body that produces your thoughts and receives thoughts from others. It is an energy field called your mental body that, like radio systems, have different frequency bands, from a lower vibrational signal to a very high signal.

The teachings of the Ageless Wisdom indicate that there are seven frequencies comprising the mental body. The lower four vibrational frequencies are grouped to form the concrete or lower mind, which, in ordinary people, belongs to the personality. The higher three mental vibrational frequencies form the abstract mind, and this belongs to the Soul and is dormant in ordinary people. In fact, the Soul's frequency and energy aura start from the abstract mental levels or frequencies. That is why when people have developed and used much of their concrete thinking, such as world thinkers and scientists, the next step is to develop and stimulate the abstract mind. The abstract mental level is where people can connect to higher creative ideas, noble principles and universal philosophies resulting in a more abstract understanding and creative expression of life.

Imagine the Earth is also made up of all-pervasive and interpenetrating mental frequencies of different grades of refinement and intelligence – like different AM and FM radio signals or short wave frequencies, and this mental field serves as the positive and negative repository of different global data, thoughts and ideas. The mind of a person is the mechanism to connect this world of thoughts and information.

The more refined, expanded and advanced the mental equipment and vibrational substance of the thinker, the higher frequencies of the world ideas he or she can tap and absorb.

The person's mental body and its substance operates like a combination of the functions of a receiver, transmitter, condenser and computer. It can receive, store, assimilate and process, then transmit, information or data. The brain is the limited hardware for the software – the mind, to use in receiving, processing and transmitting signals, which are the thoughts. The cellular phone, the TV and radio equipment exemplify the functions of the mind, especially in mental telepathy. I really admire the people who made these inventions possible. These gadgets demonstrate the initial physical stage of the power of omnipresence. Thus we can see that scientists and businesspeople are advancing their work and life purpose by demonstrating the nature and power of the Divine Creator or Macrocosm through their inventions.

The mental faculties are usually physicalized through the functions of the left and right brain. The left brain being the concrete thinking mechanism and the right brain being the subjective, abstract and creative mechanism. This is familiar to many people. But the proponents of this idea equate the mind to the brain instead of classifying the mind as the software and receiver, processor, storer and transmitter of data and the brain as the hardware or physical equipment. When your body and brain are asleep, the mind is operating on its own plane doing its own functions in the mental world. You may experience this as dreams. The brain, physical body and etheric body are in bed, but in your out-of-body experience called sleep, your emotional and mental bodies travel. When psychologists, medical researchers and scientists augment their current understanding with this idea of mind-brain relationships, then they will not spend so much time researching and focusing only on the gray matter and its chemical and electrical components and will break through with different solutions to psychological or psychiatric health problems of humanity.

I think that many health specialists in the world share the frustration of their inability to connect to higher solutions for the mental and emotional dysfunctions affecting much of our population today – without having to be limited to pharmacological and physical tools and remedies only. The psychologists have already started to unfold mental and emotional principles, but they still lack the involvement of energy strategies and inner aspects of the solutions. Once they involve the concept of the mind, emotional and vitality bodies and how to deal with their corresponding energy centers, the whole understanding and approach to mental and emotional life and diseases will experience a big change. The books on *Inner Powers* are partly written to address the need to bridge the missing links not utilized by different careers and specializations of human development, including education, psychology, business, health care, politics, religion and the culturing of civilization, with energy applications.

We will not elaborate much on the physiology and anatomy of the brain here. That is the specialization of medical science and much attention, research and products have already been developed for its benefit. Instead I will emphasize the mental powers and their development and applications for present needs and future requirements.

Here's an overall view and relationship of the mind, the energy centers and faculties of the brain as follows:

Mental-brain-thinking relations				
Mind type	Energy center in charge	Side of the brain used	Faculties used	Controlling consciousness
Concrete mind	Throat	Left	Logical, sequential reasoning and concrete thinking – in charge of knowledge and details	Personality (highest aspect)
Abstract mind	Ajna/ mid-brow	Right	Abstract, creative, subjective and philosophical thinking, general understanding and abstractions – in charge of inner principles and ideas	Soul (lowest aspect)

Power to physicalize

To study and understand deeper the inner powers of the mind, let us see how thoughts and the actualizations of reality are related, or how the mind is utilized to physicalize or affect the outcomes of events. How do your thoughts or thinking habits affect your physical body and its vitality and chemistry? Let's do several experiments. To fully unfold the essence and appreciation of this simple revelation, I suggest that you participate in the following procedures and make your own conclusions.

Experiment #1: Mind and thoughts relationship to the physical body and its chemistry

Close your eyes and visualize that you are in your kitchen. Go to the refrigerator and take out an orange. Slice it into two halves. Open your mouth and squeeze the juice into your mouth many times. Let the juice stay in your mouth. How does it taste? What is the effect on your salivary glands? Is there an effect of visualization or imagination? How did this thought affect your body and its chemistry?

If you imagine a fearful situation or negative event in your life, whether it is real or not, does it also affect your body by stimulating certain hormonal imbalances? In my own experience with clients, there is a direct effect. The mental body energy centers directly connect to and affect the corresponding energy centers of the emotional and etheric bodies. The etheric body energy centers directly control the glands, organs and body functions. Here lies the direct connection and relation of the mind, emotions, vitality and physical body. When researchers understand this, they will benefit greatly in the advancement of and future approach to their work and discoveries. This also explains why the power of the mind, expressed through techniques like affirmations and visualization, helps heal the body and emotions and increase vitality. On the other hand, a person with negative thinking habits can also devitalize their emotional and physical health through negative thinking.

What do you think? What adjustments do you need to make in your thoughts patterns to create a healthier life?

Experiment #2: The relationship of the mind, emotions and vitality

Close your eyes again and think of your enemy or a devil's advocate. If you do not have one, think of a person who hurt you deeply and focus on the event or how negative the person is. Do this for 2 minutes.

What's the effect on your emotions and vitality? Did you become devitalized or depressed and agitated? What is the effect on your feelings in the heart center at the chest? Did your stomach or solar plexus center react? What's the effect on your breathing? Did it become shallow and erratic? What's the effect on your heart beat?

With the above experiments and results, we can conclude that repeated or prolonged negative thoughts drain your vitality and negatively affect your emotions and health.

Therefore, if the mind can heal, the mind can harm.

That's why you can see the benefit of having a more refined and advanced mind. Advanced mental development externalizes as positive, powerful, cultured words and behavior, which create good relationships. Also, since contemplating on negative experiences, events or people brings us down, we should avoid it. Because mental energy is transmitted from one person to another, we are reinforcing the very qualities we are criticizing. And, finally, we are creating that negative quality and substance in our own energy field.

> You cannot throw mud at others without dirtying your own hands first.

This is also true in destructively criticizing and harboring anger towards other people even though we might be on the right side. We can avoid being a victim of people's negativity by not recycling and attracting further emotional and mental contamination, causing devitalization. I recommend that you observe this principle in your daily life to help you maximize performance and enjoy good health. Maintain a positive attitude and correct any wrong mental habits or processes as soon as possible.

Experiment # 3: The power of materialization using the mind

Do this in a standing position with your feet apart at shoulder width. Stay in this position until the experiment is done. Try to touch the floor or ground as low as you can without bending your knees. Do not over-exert! People with back, hip or knee problems should do not do this experiment. Remember how low you have extended your fingers towards the floor and how easy it was. Now, go back to the standing position and visualize that you have touched the floor or ground lower and easier than before. Repeat the visualizations 7 times. Again, try to touch the floor as low as you can, evaluate the level and feel the difference of the effort and ease involved compared to the first try. Do another set of visualization 7 times and imagine that you are even more flexible and extending your fingers even lower towards the floor. Try to physically touch the floor and compare it with your first attempts. Is there an improvement? In just a few seconds of visualizing the goal, what is the effect of your mind in achieving the goal? Did it improve your performance? Did you achieve better results with your objective?

I used to train and coach martial arts teams for competition ranging from black belt – all styles, all weight divisions – to beginner levels for adults, women and children. By teaching this mental programming to my team members and implementing the techniques many times a few weeks before competitions, we often won the team championships. We usually ended our training with a meditation and concentration to visualize the results we wanted. As a common saying says:

"What we think repeatedly of ourselves, we become."

The goals we repeatedly think about and focus on also tend to materialize faster and easier. Therefore, spend your precious time mentally visualizing or concentrating on positive end results only. If you happen to think of negative outcomes, especially accompanied by or resulting in fears and worries, determine if they are real or if your mind is reminding you of potential difficulties. This is a normal occurrence, but do not dwell on problems too long. Instead, create solutions and have contingency plans. Avoid creating goals and making decisions when you are not in a good emotional mood or if you are very angry, stressed or fearful. Why? Because the perception of what is urgent and right in this state is not necessarily correct.

When people create goals and intentions colored by negative emotions, the goal and outcomes may seem valid at the time, but in the long run, they can bring back suffering and destructive boomerangs. In creating your goals and mental intentions, always countercheck yourself by role reversals and the

Golden Rule. As you become mentally sharper and smarter, you can justify and manipulate many things as if they were correct or right. The problem is that they can result in short-term success followed by a lifetime of regret or suffering. Therefore, be careful when you create powerful thoughts, intentions and goals.

The Golden Rule

Do unto others what you want others to do unto you.
Do not do unto others what you
do not want others to do unto you.
- Confucius, Chinese philosopher

The Koran also states the following:

Noblest religion this – that thou shouldst like for others what thou likest for thyself; and what thou findest painful for thyself, hold that as painful for all others too. As ye would that men should do unto you,
do ye also to them likewise.
- The Koran, Holy book of Islam

I suggest the following to achieve long-term benefits and win-win situations:

How to program only positive end results

Visualize the positive and happy outcomes and end results of your goals and plans, including the good effects for you and the people concerned. Program by making strong intentions, wishes or visualizations that the goal will physicalize only if it is good for your whole life; if not, it should not happen. Mentally programming your positive goals needs repeated concentration and visualization to make it powerful.

Here are some procedures to successfully program your goals:

Procedures for programming and materializing your goals
1. See your goals with mental clarity and focus on the end result.
2. Mentally see your goals already accomplished and visualize only the end result.
3. Have high expectation and strong conviction that the goal will succeed.
4. Work on your "luck": Intelligent hard work + good karma

Procedures for programming and materializing goals

Step 1: Set your goals with mental clarity and focus on the positive end results.

This is the first important step without which the other steps have no power. Very successful people and achievers are masters of this first step. Despite obstacles, they achieve their goals because from start to finish, they are very clear about what they want. Remember that indecisiveness is a vice. Just make sure the goal is in accordance with your life's vision and purpose and is aligned to your principles. Once the goal is set, the next step is to maintain focus and constancy.

Step 2: See your goal already done successfully and visualize only the end result repeatedly, not the process of achieving it.

The more you concentrate on the end result, the stronger it becomes.

Step 3: Bring high expectation and strong conviction that the goal will happen successfully.

This step creates strong enthusiasm and energy for the goal.

Step 4: Commitment

Commitment is very important. All the above steps will not yield the speed and power of materialization without real commitment. Commitment is the power that drives you to finish the goal with determination and constancy of effort. Part of commitment is the creation and implementation of schedules and deadlines. This is the backbone for your end results.

The Power of Knowing Your Goals Clearly: The Sam Walton Story

Sam Walton, the founder of Wal-Mart, one of the most successful businesses in the US, is a great example of the power of goal-setting. When he started to envision Wal-Mart's strategies in his mind, he probably committed to being the best in his line of business. He was determined to adapt the best processes of retail stores and innovate them to be better and to bring the lowest prices to smaller US cities.

According to John Huey, who wrote about Sam Walton in the *Time* 100 Magazine, "... in 1985, *Forbes* magazine determined that his 39% ownership of Wal-Mart's stock made him the richest man in America."[2] When Sam Walton died in 1992, with a family net worth approaching $25 billion, he left behind a broad and important legacy in American business, as well as a corporate movement: As Huey recounts, "In 1998, Wal-Mart was the No. 4 company in the *Fortune*

500, behind only General Motors, Ford and Exxon, with annual sales close to $120 billon."[3]

Wal-Mart's Board of Directors and staff (and Sam Walton before he died) are always clear on Wal-Mart's main goal, enumerated in their slogan: WE SELL FOR LESS ALWAYS. This focus is what makes them very successful.

If you know the positive end result and constantly focus on it and maintain physical, emotional and mental commitment to it, almost nothing can stop you from accomplishing it. It is just a matter of time.

Step 5: "Luck"

There is an expression that luck goes to the people who work for it. What does this statement mean? Luck is the good grace of entitlement, which has two forms, current and ancient.

- **Entitlement in this life**: Many people start at the bottom but today are at the top of their chosen careers. I cited Oprah Winfrey, a television personality in the United States, as an earlier example. According to an article written by Deborah Tannen, she started out as a simple girl with no special lineage. Born on January 29, 1954 into a poor family, she had amassed an estimated personal fortune of more than $500 million by the time she was 44.[4] I believe that her generosity and her philanthropic works supporting educational funds to send disadvantaged kids to college and many unpublicized charity works have given her the grace of good luck (in addition to her hard work and ability to invest properly). As Deborah Tannen went on to write, "Winfrey stands as a beacon, not only in the words of media and entertainment but in the larger realms of public discourse."[5]

A Christian teaching says:
"God is not mocked:
For whatever a man soweth, that shall he also reap."
- Galatians 6:7, The Bible

This universal law is also mentioned in most religions including Islam, Buddhism, Hinduism and Judaism.

- **Ancient endowment**: If religions believe that what you plant today, you will harvest in the future, planting and harvesting must have an interval period and process. Many believe in rebirth; others don't. My question is: How do we explain why some people have an easier time and effort, amass wealth and get what they need without working very hard for it? Some people are not mentally gifted, were not born into a fortunate lineage and are not necessarily spiritual, but they suddenly become instant millionaires by winning the jackpot lottery. Other people are not very educated, do not work hard in their life, are not good looking, but suddenly good luck comes and they marry a wealthy person. This person dies after a few years and the spouse is left with sudden wealth. You might say it is coincidence or good luck. But I think there must be an ancient source of invisible wealth endowment. Is it a coincidence, or is there a perfect spiritual karmic mathematics, which looks for people who worked for what they receive (maybe not in this life, perhaps in ancient lives) and were very generous and benevolent personalities who helped many people. Perhaps now is the time to harvest what they planted in the past.

Step 6: Intelligent hard work

Hard work guided by proper organization, time management and creativity, lubricated by good karma or entitlement to receive, is called intelligent hard work. Herein lies the significant difference between mentally polarized people and emotionally reactive people. Mental personalities are usually more structured whereas physical or more emotional people are more impractical and unfocused. Many successful leaders and achievers have done so well and have so effectively materialized their goals because of proper short- and long-term preparation. This expresses itself as the right timing of events, good investments and good outcomes of decisions, which make a person, organization or country appear lucky. The success of prosperous or developed countries or organizations did not happen overnight. They worked hard, sacrificed and prepared for the next harvest of abundance and success.

Intelligent hard work is a product of a well-organized, well-informed and disciplined mind. In this epoch of our civilization, I predict that the countries and the cultures that continually evolve inner powers of the mind will be ahead in politics, education, business, arts, sciences, technology, culture and spiritual development, especially if these nations and peoples maintain good will, will-to-do-good, compassion and generosity in other words the powers of their hearts. They will be unbeatable, and they will be the ideal model of future culture – a culture guided by love, wisdom and science. This will be the result of using mental powers for greater service and for the advancement of humanity and the world.

As Winston Churchill stated:

> "The empires of the future will be the empires of the mind."
>
> - Winston Churchill, British Prime Minister

Ramifications of mental power

Elements of Mental Power
Common sense
High grade of common sense
The concrete mind
The abstract mind

Elements of mental power

First element of mental power: Common sense

In normal human beings, common sense is a body of reasoning using knowledge or feedback through the five senses. The data is processed by the concrete mind or by the emotional instinct. Therefore, the reliability of the conclusion still depends on the intelligence of the mind and emotions to interpret the perceived data. Because common sense generally depends on old conclusions, history or existing data, there is a danger of tapping into the Earth's library of mixed truth and false information. There are perceptions based on false data, which appear seemingly true and are justified as correct because of the "curtains" that screen negative old programming or beliefs. Also, the emotional and mental bodies and chakras used to interpret and screen data from the mass consciousness of humanity are immersed in or connected to many oceans of existing thoughts, ideas and emotional programs. This is why for a person with good discrimination, common sense is very useful, but for someone who is very emotionally impulsive and non-discriminating, their common sense is not reliable.

To give you a bigger picture, let's see how the library of humanity's common sense is developed and what generally comprises the world of common sense:

Elements of Mass Common Sense
Ideas and knowledge from previous cultures, civilizations and races
Knowledge from science
Teachings from education and religion
Beliefs and superstitions of any kind from all sources
Illusions of thinkers and groups
Media information: true, false or partially correct
Experience from upbringing and daily life

Every day, this body and source of common sense is building up through the aggregate of humanity's thoughts and emotions, especially now that we have global connectedness and Internet use. The problem arises when these mental and emotional programs and powerful streams of information called mass common sense feed people's minds with wrong input. This will lead to wrong conclusions and wrong decisions that make human beings suffer and fail. Therefore, a guideline is needed for the proper use of common sense:

Guidelines for the Proper Use of Common Sense
Be open-minded, not fanatical in believing all the things you read, hear or see.
Be careful with gossip and news.
Use scientific data or opinions from experts as much as possible.
Do not rush to conclusions if you are in doubt.
Use reliable historical data to validate your common sense.
Seek advice from a consultant or expert.
Only study good books or materials from reliable sources.
Surround yourself with brilliant and successful people or look for such a group.
Approach things with a more universal sense of service.

Be detached from your usual likes and dislikes.
Know your weaknesses and strengths.
Values and virtues are helpful in situations where your reason is doubtful.
Be realistic.

Let's examine each of them in detail:

- Be open-minded, but not fanatical in believing all the things you read, hear or see. Always ask questions regarding the what, why and how of anything. It is helpful to see the bigger picture rather than believing everything.
- Be careful with gossip and news, either from the media or personal conversation. Anything not useful to your life, which brings separativeness or failure, should be cast aside or dismissed. Validate information and investigate the truth if you are concerned with the given information or news.
- Use scientific data or opinions from experts as much as possible.
- Do not rush to conclusions if you are in doubt. Assess your common sense with your hunches or physical and emotional instincts.
- Use reliable historical data to validate your common sense.
- Seek advice from a consultant or expert on the subject matter to avoid unnecessary mistakes. Get a second or third opinion if in doubt.
- Only study good books or materials from a variety of reliable sources to avoid admitting more junk or false data into your mental files or emotional memory and to be well-rounded in different subjects.

- Surround yourself with brilliant and successful people or look for such a group to associate yourself with. Good common sense is contagious through group interaction.
- Since your perception is based on your emotional and mental analysis, which has preferences or biases, when confronted with conflicting information, approach things with a more universal sense.
- Be detached from your usual likes and dislikes if you are deciding for a group. Remove biases by prioritizing what is best for your group or team.
- Know your weaknesses and strengths. When making decisions, know where your common sense or reasoning is coming from, and be fair and realistic. Do not allow decision-making and reason to be colored by old sentiments, personal grievances or negative motives. Use role reversal to test for win-win situations.
- Values and virtues are helpful in situations where your reasoning is doubtful. Applying the universal values of life guarantees that you are right even if your common sense dictates differently.
- Be realistic. Know the real facts and recognize what is practical and objective. Shift to mental reason when you become too emotional or sentimental.

The power of common sense depends on the state of the emotional and mental bodies and on the degree to which spiritual values are developed. Therefore, a lifetime of building character and culturing the emotions and the mind is the reliable path to mastery of common sense for normal people.

Advanced faculties supersede normal reasoning in advanced people. Developing the abstract mind is the next step for most of humanity in order to transcend the illusion and confusion of mass thought. Therefore, humanity will progress to the next stage of mental development, the abstract mind.

Second element of mental power: High grade of common sense

We've discussed common sense, but there's an even more powerful mental faculty associated with common sense: a high grade of common sense. This is common sense that has been upgraded to a more elevated knowing. More learned human beings – not necessarily degree holders or those educated in schools – born with a better apprehension of life from the time they were young are equipped with this mental faculty. One factor that makes them better thinkers is an extra faculty of inner sensitivity that allows them to experience truth faster than most people. They also have greater exposure to and interaction with intelligent and wise people. These intelligent and wise individuals may be their parents, spiritual teachers, coaches or mentors. Some who become great personalities have been exposed to great and successful people and leaders, so through energetic osmosis and positive contamination of qualities, they acquire a higher grade of common sense and intelligence. Also, by observation of smart people and parents, children become more intelligent. In fact, some people with this faculty do not have degrees or a formal academic education. And some drop out of school because they are bored or prefer to learn differently.

A High Grade of Common Sense: The Bill Gates Story

Bill Gates dropped out of Harvard to co-found Microsoft in 1975. He continues to head up that company and has been assessed as the world's richest man. According to David Gelernter who wrote for *Time*, "Gates is the Bing Crosby of American technology, borrowing a tune here and a tune there and turning them all into great boffo hits..."[6]

Gelernter continues: "The Gates Road to Wealth is still a one-laner, and traffic is limited. But the idea that a successful corporation should enrich not merely its executives and stockholders, but also a fair number of ordinary line employees...is potentially revolutionary."[7] Many accuse Bill Gates of monopoly. Others see him as an idol. But one thing I sense is that he has a high grade of common sense that many very educated people lack.

I have observed in many of my clients, friends and acquaintances that a high grade of common sense can be developed by international travel and especially by living in different cultures. Even with my intense training from my early spiritual mentors and deep study of advanced books from the Ageless Wisdom, without having lived and worked in Asia, the Middle East, India, North, South and Central America and Europe and with constant travels to many countries teaching, learning, experiencing and merging with the different cultures, my knowledge and common sense about religions and cultures would have been more theoretical and idealistic. Being in the energy field of these different places, you gain a deeper understanding of why we exist today as such – and harmony through contrast and unity through diversity become wise ideas.

In addition to exposure to various cultures and having a good mentor, the opportunity to mingle or be working with brilliant people, especially leaders, is one of the fastest ways to acquire a high grade of common sense. Most highly intelligent leaders have this virtue. In fact, a high grade of common sense has to be progressively and continuously developed and refined. For the intelligentsia, this one is of their most advanced faculties to develop.

Third element of mental power: Concrete mind

Scientists (especially physicists), lawyers, engineers and accountants are experts in concrete thinking power. The detailed and organized search for truth is active in the throat chakra and the left-brain process, although the more advanced scientists and engineers start to incorporate the abstract mind and right brain processes. In ordinary people, the concrete mind is starting to develop, most of our education systems today still focus on concrete mind development. Our schools tend to emphasize memorization and a deductive learning style. Even our world history is taught through memorization instead of understanding and abstraction of human lessons in order to recognize which mistakes not to repeat in the future.

Many advanced children are bored in school. Sometimes they are classified as attention deficit children. While some children classified as ADD or ADHD do have difficulty processing information, others are reacting to the ordinary, slow, linear learning environment. In fact, their minds may be faster than their age group and they need to occupy some mental gaps by doing something else or by thinking on a different frequency and with different intentions. I have child and adult clients who are labeled ADD but who have developed minds. I also had classmates who excelled in math, but performed poorly in music and arts. They are very left-brain people. People who are sometimes called geeks and nerds seem to have partially developed very good concrete mental faculties, but lack emotional body development. That's why they can be geniuses with concrete mental expertise, but without the ability to relate to others emotionally.

How does the concrete mind work and gain its power? Again, let us enumerate the process:

The concrete mind creates thoughts and ideas usually deduced from common sense information or through a visual, sequential thinking process. Let's look at some examples and comparisons:

Process	Created	Creating Body
Think of a green apple.	Apple energy thoughtform	Concrete mental body
Have a desire to eat it.	Emotional appetite called desire	Emotional body
Recall a successful event.	Reconstruction or revitalization of that event	Concrete mental body
Experience and feel the event.	Emotional feeling about the event	Predominantly emotional body
Visualize and think of your brain	Visuals and thoughts about the brain	Concrete mental body (not the brain)

People from different cultures with different languages create energetic thoughtforms with the same forms and features. For example, a North American says and mentally thinks "apple" and a South American says and mentally thinks "*manzana*", but the mental substance created by both has the same energetic form. This is the basis of mind reading and telepathy. Another important point is that thoughts are not feelings and vice-versa.

Thoughts come from the mind, and feelings emanate from the emotional body.

The substance forming thoughts is more refined than the emotional or feeling substance.

Functions & Powers of the Mind
Information storage & memory
Organizing, problem-solving & decision-making
Power of research & investigation
Concretizing ideas
Mental power of abstraction
Power of imagination
Creative power

How the mind works: The concrete mind memorizes & stores information

The brain does not store data. It is the concrete mind that absorbs and stores information. The brain is the hardware used to process information when data is first registered. If you recall information, you draw from the data bank, which is the concrete mental body.

There are energy compartments or memory banks in the concrete mental body that serve as repositories of information or data. The grosser the thought, the lower the frequency at which it is stored. The more refined the idea, the higher the frequency at which it will be assimilated. I call these "mental diskettes" or compartments. Therefore, advanced training given to my students uses compartmentalizing the mind techniques to improve higher mental powers, faculties and discrimination. The mental body and its energy substance need purifying and decongesting. It's like your data storage unit in your computer. You need to purge unnecessary or obsolete data or documents to the recycling bin or trash. There is also obsolete data from your younger years and previous training that have to be erased or purged when they are irrelevant or obsolete. Then the new data and updated valid information should replace or supersede the old obsolete information. This complete method of compartmentalizing the mind will be discussed in detail in Volume II of *Inner Powers.*

People become so confused with conflicting information in their decision-making because their childhood or educational data from fairy tales, fables, parables and early conclusions about life are still intact and competing with the more realistic adult interpretation of life. This is the internal tension in people who grapple with conflicting ideas of archetypal models of truth and their own repeated valid experience. They don't know how to behave in expressing the higher truth currently perceived because of their great guilt and fear from religious beliefs or the influence of the old learning pattern. This is true with sexuality, religion and husband-wife psychology. Many of my female clients experience this predicament – conflict among religion, cultural and parental upbringing, which are totally opposite from their current priorities and preferences. Mental programs and memories should be upgraded and updated in order to resolve these conflicts.

Memory acts as powerful data for the concrete mind and common sense. The faster and better you can memorize, the more data upon which to base decisions and deductions. The only problem is: What kind of data are you using? As we say, "Garbage in, garbage out." That's the danger, so we should be selective in what we read, learn and put into our memory. This includes visuals and input from TV, news, books, audio and video references. This process of learning affects children, teenagers and adults today. They sometimes pick up the wrong social models, TV programs and music selections. This is the danger of fast memorization techniques used in speed reading. The untrue information that we have quickly mentally photographed in the past might be stored without discrimination or a process of reason. The materials and information that feed our senses and memory should be screened properly. There is such a thing as mental diet and proper mental nutrition.

How the mind works: Organizing, problem-solving and decision-making aptitude and powers

The design of the concrete mental faculty is to construct ideas, deduce and process details and organize them into the mental body. This is expressed as the ability to use details and particulars in the general scheme of things. Again, this faculty is expressed through the development of the throat energy center, especially in the mental body.

If the mental thoughtforms and goals in life are clear and strongly vitalized, and the mind is equipped with sufficient information related to the project or subject, the mental aura is bright, calm and strong – the organizing and concretizing ability and power of the mind is efficient. Best results are achieved when ideas, goals, procedures and schedules are properly and clearly written. Sometimes it's better to rewrite them several times until everything is simplified, clear and organized. As you write your thoughts and ideas, you may notice that before your started writing, you assumed you knew everything clearly already, but the moment you try to write your ideas and goals on paper, you pause periodically and change the sentence, then sometimes cross the statement out. It means that you are not as clear yet on what you really want, or your mind has not organized it yet. So writing your thoughts and ideas helps you clarify, improve and organize them.

In problem-solving and decision-making, the mind works best when you write the following:

1. What is the real issue or problem?
2. Is it real or a byproduct of fear or reactive emotions?
3. Who is involved and affected by the problem?

4. How much time do I have to solve it?
5. Who can help me, and what are the necessary talents or aptitudes required of them?
6. How much money (if required) is involved to solve the problem, and what is the source of funds?
7. What are the probable solutions? Select and analyze more than 2 possibilities. What are the benefits and consequences of each solution?
8. After using your mental evaluation, countercheck with your emotional and physical instincts to determine the best solutions and decisions. Ask for an expert's opinion, if required. Integrate solutions for win-win outcomes. Use role reversal to test fairness to people involved.
9. List the best solution, procedure, budget if required and personnel to help, if needed.
10. Make the complete schedule and final plan, including contingency plans for potential obstacles.

Once you write and evaluate everything, mental confusion is minimized and the overwhelming effect of emotional reactions is reduced or eliminated.

Writing things and organizing them on paper exposes the erroneous blind spots and unforeseen obstacles.

It also shifts you from the negative side effects of emotionalism into a more mentally intelligent approach to managing crises or creating plans. This is the next step for masses of humanity – to employ the mind to solve its world problems.

How the mind works: Power of research & investigation

The concrete mind has this special faculty of comparing duality and the power to analyze interrelationships of data and information. By continuous focus, data-gathering and investigation of deeper and broader implications, applications and details, the mind will be able to penetrate the greater reality of forms…just like drilling down into the nature of the physical body – from the general structure to the organs, tissues, cells, atoms, molecules and sub-molecular particles. Then the next step in investigation goes beyond the physical form to the world of invisible energy.

This unending investigation using the concrete mind Is well demonstrated by the modern scientist, Harlow Shapley, the first person to measure the size of the Milky Way Galaxy as 300,000 light years across.[8] He claimed that the Milky Way Galaxy was the cosmic whole. The story of this great scientist was challenged by the new discoveries of Edwin Hubble, a famous astronomer who used the 100-inch/25-centimeter long Hooker telescope at the Mount Wilson Observatory. He discovered that the universe extends beyond the edges of the Milky Way Galaxy and in 1929, he further concluded that the universe was expanding.[9]

This discovery eliminated Einstein's uncertainty about his general theory of relativity. Earlier astronomers assured Einstein that the universe was neither expanding nor contracting. These conclusions affected his earlier equations, but by the sheer power of timing, the potential blunders in his equations were resolved by Hubble's discoveries into what we know as Einstein's equation of relativity.[10] So, you see, the concrete mind power of research and investigation can lead to great unfoldment of valuable tools and equations, but those who don't use them properly or validate them with other methods can be victims of wrong and incomplete mental evaluation. The power of the mind is a double-edged sword.

How the mind works: The mental power to concretize ideas

If you observe and study how ideas, technology or science are physicalized into objective reality or products and services, the general patterns used are as follows:

Type 1: Mental power to concretize ideas, products or services

1. First, the idea is known.
2. The idea is studied and broken down into principles.
3. The principles are organized and broken down into processes and procedures.
4. The processes and procedures are subdivided and broken down into methods and techniques.
5. The methods and techniques are broken down and organized into steps.
6. The steps are either qualified, quantified or timed.
7. The whole system and technology are manufactured, priced, packaged and sold.
8. Marketing, continued research and product development ensures the innovation, improvement and new designs in technology, products or services.

The above is a display of how concretizing power of the concrete mind works, creates and innovates. When I see successful organizations and technology, I know that many great minds have pooled their concretizing mental powers together to make the endeavor successful. Successful organizations or endeavors are not like mushrooms growing overnight. They are a byproduct of tremendous mental power and effort.

Type 2: Mental power to organize life and manage time

The power of the concrete mind can also be used in many ways to organize and concretize your life's goals – and manage your time properly. Here's how:

1. Think of the most important areas of your life that you need to fulfill.
2. Identify the most important vision, mission, plans and goals that you need to achieve, short- and long-term.
3. Identify and properly design the most important roles corresponding to the mission/goals you have set for yourself.
4. Knowing your roles and goals, identify the most important activity and job description for you to perform in all the chosen areas of your life.
5. Knowing the most important activity to complete monthly, weekly and daily, design the most important events and results you need to implement, qualitatively and quantitatively.
6. Create a schedule in your calendar or planner to list and organize all the details.

The most important initial step of organizing your life and managing your time properly is to know exactly what you want to do in life and what you want to achieve. This is summarized in your vision and mission, which will become a plan.

> The vision, mission and plans of your life are the lighthouse, compass and map that you need to reach the shores of success and the land of fulfillment.

Organizing and planning your life and managing your time wisely do not only depend on how disciplined you are about putting deadlines in your planner or calendar. It is more than just writing, following and beating deadlines. The schedules and results should constantly follow a path of coherence to your life's purpose. That is why I am suggesting that:

> Your true time and life management revolve around priorities based on your noble vision, practical mission and realistic plans.

For those who want to use the mental powers scientifically and technically, you can maximize performance and success by employing them to organize your time and life.

Type 3: Mental power to physicalize goals

1. First, define the goals, project objectives and end results.
2. Determine and classify the general activities.
3. Subdivide the activities into tasks, functions or roles.
4. Break down the tasks into work or job descriptions.
5. Establish a scope of work and results to be achieved, with their quality, quantity and rhythm of timing.
6. Determine the financial and material resources and their sources.

7. Determine and organize the implementers and workforce with the top leaders as a priority. Once they are chosen, they can help with the details of organizing. Make sure the personnel expenses can be covered in the financial budget, otherwise adjust the plan.
8. Determine possible problems and obstacles. Make a contingency plan for potential problems ahead of time, if possible. Try to avoid surprises.
9. Create the final schedules with deadlines, follow-up dates and a list of implementers and targets.
10. Regularly monitor and evaluate the plan, situation and project development. Adjust the plan as required.
11. Always know the greatest benefits and service of achieving your goals. This will inspire you to make the required sacrifices.

Now, if you observe the sequence of this concretizing technique, it involves a flow from the general to the particulars. Most great projects and goals are physicalized with the power of the mind and constant focus. When I started to write this book, I had already seen and organized the general contents and chapters in my mind. I already saw the whole outline from the beginning to the end. I had already imagined the flow of ideas. I knew what I needed to write, and I just started to scribble all the words and ideas, letting them flow according to the pre-determined goal of the book.

No matter how sharp a person's mind, it can still be honed to become even more extraordinarily sharp and brilliant. There is no end to its training and usefulness. That's why many corporations are just now investing a lot more in the proper training of their workforce, especially the management. The new system of upgrading the workforce includes mental training to think of new ways to create, not just to follow orders or manage linear models of business execution. I wish that more corporations would realize that the true investment and assets in the organization are the people…and that the real, tangible assets are the quality of people in the workforce and the mental powers of their leadership. Technology easily depreciates, but people can appreciate for higher-value investments – with proper mental training. I believe that the real job of an executive, professional and especially a leader is to be the real thinker and mental powerhouse of the organization..

Fourth element of mental power: Abstract mental power

So we've discussed the functions and faculties of the concrete mind. What comes next? Abstract mental power. Abstract, subjective, philosophical, principle-based thinking belongs to the abstract mind, which is the lowest vibrational body and faculty of the Soul. Therefore, when people start using the abstract mental faculties, they start utilizing the power of the Soul. The new modern generations, especially the incoming children, are more equipped with the power of the abstract mind. You can observe how many more advanced children learn new things. They start asking you questions starting with, "What is this?" and "How is it?" and usually end with, "Why does this exist?" The questions of what and how are an exercise of the concrete mind – to accumulate knowledge and express the idea practically. When the point of interest is the why's of life and things, then the abstract mental faculty is being stimulated.

Another training tool for this higher mental faculty is asking questions that have not been experienced or known by the concrete mind or which have no logical meaning. For instance, in Zen training, the enlightened teacher called a Roshi gives coans or puzzles that you need to contemplate for a long period of time until you have an appropriate answer. I did not say a correct answer because it is not really the concrete accuracy of the mind that has the answer but the abstract faculty to discover the answer. For example, here's a familiar Zen coan: What is the sound of one hand clapping? Try using the reasoning of the concrete mind: The memory does not recognize any experience, and it cannot comprehend with deductions unexperienced non-linear concepts, so it relinquishes the thinking power to the abstract mind, which will search for the philosophical answer, concept or principle. So the answer is not always a straightforward statement.

Another difference between the concrete mind and the abstract mind is that the concrete mind uses a focusing and concentrating technique while the abstract mind uses the techniques of awareness and understanding faculties. When you say, "I know", you are using the concrete mind, but when you say, "I understand", you are using the abstract mind. For Western culture, there's a similar abstract mind-training puzzle: Which came first, the chicken or the egg? Ponder on this and observe your mental faculties and processes.

Now that you know the basic differences and functions of the concrete and abstract mind, we can proceed to the practical strategies: How do you stimulate or fertilize the abstract mental powers?

Training for the abstract mind

1. As much as possible, investigate the "why's" or purpose of things, events, life, existence, and uncommon knowledge. If you keep on answering the why's, you enter into the realm of principles and concepts, which in the long run leads to universal ideas and laws. Then you are in the mental frequency of the abstract world or energies.

   Withdraw periodically from routines and repeated activities. With redundant activity and events, the abstract mind is not exercised because it is your physical instincts and concrete mind that are employed daily. Look for new assignments or add activities to your schedule that are unfamiliar to deprogram the rut in your mental functions and execution.

2. Study new and advanced technological and scientific discoveries and ask yourself, what are the new principles and concepts behind them? Look for the creative and innovative ideas involved and study them as universal principles and subjective laws. For example, I interpret the invention of the TV, radio and cellular phone as a Divine manifestation of Omnipresence on the physical level. When I observe the Internet now used globally and packed with huge libraries of information, I think of the abstract idea of Divine Omniscience. When I think of the nuclear or atomic bomb, I see a crude material expression of the Omnipotence of creation. When I hear of new discoveries concerning procreation and artificial reproduction in vitro, including production of tissues from stem cells, I interpret it as humanity evolving to become a physical creator. So with the cloning of animals and species - these are physicalized powers of the human mind primarily using the abstract mind, which then physicalizes by the power of the concrete mind. These types of mental practice train the mind to understand and relate concrete discoveries to universal principles.

When I read or hear about important political or financial events in the news, I try to penetrate the underlying causes and effects, especially the hidden meaning of those events. A real thinker knows a lot more than common sense and trains himself or herself not to be a human parrot. When humanity employs this practical continuous method of mental training, starting from childhood to the post-graduate education level, we will experience a generational leap of consciousness from the mundane mediocre focus on the trivial events to deep understanding of the world of causes. Most thinking processes have evaluated world events only at the surface level, always blaming somebody for one's sufferings or mistakes. Once the abstract mental principles are unfolded, the search for solutions, missing gaps and links for the meaning of our existence will be directed from external to internal. Then the real truth and causes will be understood. World history will be an important teacher, and wars will be avoided. True peace and unity will be an inevitable result.

3. Study poetry, quotes and literary works from great philosophers, Sages and writers like Lao Tzu, Shakespeare, Tagore, Rumi, Hellen Keller and other excellent artists.

William Blake is a very good example with his famous lines:

> To see a World in a Grain of Sand
> And a Heaven in a Wild Flower
> Hold Infinity in the palm of your hand
> And Eternity in an hour.[11]
> - William Blake, English poet

Penetrate the underlying principle and universal values and concepts in their poems, quotes or literary works. Many of these writers experienced advanced abstractions of material life and understood some inner cause and underlying reality that can explain mysteries and inner experiences. Once you realize and understand life's subjective principles, it exercises and fertilizes the abstract mind. That explains why the more advanced modern scientists like Albert Einstein, Edwin Hubble, Thomas Edison and Madame Curie were not just utilizing their concrete mind, but were also able to connect with universal principles and concepts through their abstract mental faculties.

If you study the movie industry, some of the script writers and directors are geniuses in their own right, producing and directing what they consider works of arts and creativity, but which actually involve the use of concrete and abstract mental faculties. A good example is Stephen Spielberg, who directed *ET* and *Close Encounters of the Third Kind*, both interesting abstract movies.

One of the best ways to train and develop the abstract mind is reading and studying the books of an Advanced Sage, Holy Master Djwhal Khul as written by Alice A. Bailey. When I first read these books, I could hardly finish 10 pages an hour. The problem is not the speed of reading but how to comprehend the real meaning of the teachings. It took me a few months to become familiar with the esoteric terminology. Then, with persistence, I was able to follow through and understand most of the essence of this monumental body of esoteric work.

When the concrete mind does not comprehend the meaning, the personality surrenders. Then the Soul, using the abstract mind, takes over. After that, the higher meanings and deeper essence of the abstract writings are revealed and understood. It takes commitment, patience and persistence, plus inner power development techniques, to build the energetic muscles of the abstract mental body. Truly, the books of this Great Sage have served as one of my teachers and guides in more advanced esoteric work. They bring an inspiration that helps synthesize the entire body of teachings and practical training I've learned from all my other earlier mentors. It would be difficult, if not impossible, for me to write these books without the previous training of Holy Master Djwhal Khul's teachings.

4. In problem-solving, experience a different strategy from the usual emotional reaction and instinctual solutions. Try to understand the problem first, by asking:
    a. What is the lesson behind the problem for me and the people concerned? What was the violated virtue or principle? Why has it happened?
    b. What are the adjustments I need to make to avoid having it happen again?
    c. What are the principles and values gained if I solve it, and how do I apply them immediately? Realizing the principle and concept without making a repair is not complete. Learn by avoiding the mistake and resolving to correct it the situation, if required.
    d. Take a few hours of silence every week undisturbed by anyone just to think and create new ideas or abstract principles and concepts that are forward-looking or futuristic. If you are the leader, you should always be the one who has more advanced and futuristic ideas and information than others. Respect for leaders often comes from the fact that people naturally respect people who

know more than they do. That is one of the jobs of a powerful thinker...to be thinking ahead where no one has gone before. So you can use some weekly questions to improve your work and performance in life:

i. What new principles and ideas can produce extraordinary improvement in my ability to deliver results in support of my goals and service? Above-average thinkers should do 1 hour of pure thinking weekly. Super-achievers or leaders of leaders should do 2 hours per week.

ii. What is the future of this business, project or role? What concepts, principles or ideas do I need to employ for explosive growth?

iii. Review and internalize your life's vision, mission and general plans more deeply. There are always hidden meanings and untouched depths of your life's purpose. Unleash and unfold them in a more refined, noble and cultured way.

5. Always focus on breakthroughs or innovations that most of humanity needs and which have not been seen before.
6. Always search for even greater awareness of the purpose of your life and higher existence. Who are you? Where do we come from? Where are we going? Sometimes you need to start from zero, or from what the Zen teachers call emptiness, to really search for the right answers. As you search for deeper universal truths, you can partially connect to Macrocosmic purpose using the abstract mind. This process is like opening a spiritual antenna for higher reception of ideas.

How the mind works: Power of innovation & creativity

What's the difference between innovation and creativity? It is best expressed in this statement:

> **Innovation is converting the dirt road into blacktop while creativity is constructing a new superhighway.**

While innovation is most employed in revising, adding, subtracting and converting old created things into newer better forms and features, creativity is a new idea that dramatically changes an old idea or creates a new feature or form. We can see some basic examples in the improvements in the design and features of some familiar products or technology:

1. *Car Manufacture:*

   Innovation:

   - Add an offroad 4-wheel drive feature to 2-wheel drive for snow and muddy road applications
   - Convert a 4-cylinder pickup to a 8-cyclinder truck for more power

   Creativity:

   - Add an inexpensive hubcap to improve the appearance of the wheel
   - Add a cup holder to both front and back passenger areas in a car. (This may be one of the single most creative additions to cars in a several years.) Why didn't anyone in the automobile industry think of these creative yet inexpensive additional features earlier? The smart person who thought of cup holders used their creative powers not only to innovate but to plan a new concept.
   - Mini-vans and SUVs to meet specific family needs not addressed by traditional cars or trucks

2. *Laser pointers:*

Innovation

- Improving an old model plastic pointer for teaching into a stainless steel collapsible pointer

Creativity

- Inventing the laser pointer (pioneering)

Innovation

- After a few years, laser pointers were improved with the addition of lines and different highlight shapes in the original laser pointer (an improvement)

3. *Razor Blades:*

Innovation

- Improving the one-sided razor blade to double-sided razorblade for shaving

Creativity

- Invention of single-blade disposable razors (pioneering new idea)

Innovation

- Improving the single razor into 2- and 3-blade types

Creativity

- Inventing stick-and-pull adhesive for hair removal (more so for face, leg and arm hair)

4. *Projectors:*

Innovation

- Improving the capacity of the lamp brightness and making the transparency projector collapsible into a lighter, small hand carry size

Creativity

- Inventing the LCD or video beam projector that projects an electronic document from a computer (pioneering)

Innovation

- Improving the LCD into a lighter and smaller size with a remote control and easy hook-up

5. *Rockets:*

Creativity

- According to the writings of Jeffery Kluger[12], Dr. Robert Goddard invented the first supersonic and multi-stage rocket with fin-guided steering in 1935.

Innovation

- German scientists visiting, investigating and inspecting Goddard's rocket before the war developed their allegedly copied V-2 missile bombers used during World War II.

Innovation uses concrete mental power to revise, rectify or convert old things to better forms and improvement. It still involves the concrete mind and partially the abstract mind when there is a dramatic change of concept. Creativity is not revision or just mere addition. It is not just an artistic faculty. It is a new pioneering principle, concept, idea or activity.

Creativity is exemplified by Walt Disney. According to Richard Schickel, Walt Disney, the founder and powerful creator behind the Walt Disney entertainment world. He invented his pioneering work, Mickey Mouse, the first talking cartoon (1928), Snow White, the first full-length animated film (1937), Disneyland (1955) and conceiving Epcot (1964) – created sensational breakthroughs and pioneering work in the entertainment and movie industry.[13] The greatness of his personal growth came from his decision to flee from home at the age of 16 after a difficult childhood of poverty in order to become one of the world's most influential animation pioneers of our time.

If the abstract and concrete mind are activated, integrated and employed to discover and create new tools for the evolution of life, greatness can be achieved and success is predictable. The stories of the lives of successful personalities and celebrities in this book are employed as case studies to further elucidate the principles of inner powers and the process of how these faculties derive from the mind or emotions. The stories do not imply that I approve or disapprove of their lifestyle or personal issues.

Creativity is best expressed by this statement:

> Do not go where a path may lead you,
> go instead where there is no path and leave a trail.
> - Ralph Waldo Emerson, American poet & essayist

How the mind works:
The power of imagination as a faculty of the abstract mind

The creativity of the abstract mind is never complete without mentioning the power of imagination. I usually call it creative imagination. There are two types of creative imagination: imagination of the mundane and imagination of higher creative forms towards the future. Most children and "spacey" adults enjoy the first type, which we call daydreaming. This may lead a person to get out of body and experience inner travel to another plane. It is usually the imagination of the normal world. It is like your experience when sleeping.

The real power of imagination is the second type, which many great artists and some fiction writers use. This realm is a frequency or state of consciousness of the abstract mental world where ideas and futuristic concepts are found. That's what William Blake, one of our abstract-minded artists, meant in his statement:

> The man who never in his mind of thoughts
> travel'd to heaven is no artist.[14]
> - William Blake, English poet

When a person's abstract mind is already active, it can easily connect to the Earth's abstract realm through meditation, creative imagination or while sleeping, when people get out of body. The Earth's abstract dimension contains the world of ideas and principles not usually known to the concrete mind. When a person's mental body is not rigid and is refined and playful, it is easy to tap into this higher realm.

What is the difference between visualization and imagination? They are both apparatuses of the mind, but visualization is a faculty of the concrete mind and imagination is the faculty of the abstract mind. That's why creative artists are more mentally fluidic compared to an engineer whose mind is more fixed and objective. Innovation utilizes visualization more and creativity uses the imagination more.

Let's study several examples of these two faculties:

Example 1: Visualization versus imagination

1. You can visualize your fingers because you have seen them before. This process uses the concrete mind.
2. You have to imagine how to clap with only one hand because you have not seen that at all. The concrete mind ceases to function, and the abstract mind starts to be stimulated.

Example 2: Visualization versus imagination

1. You visualize when thinking of when you were young: What were your favorite toys? This technique involves the concrete mind because you travel to your known past.

2. You imagine when you envision your life 10 years from now: What will you be wearing and what will you look like? This mental process utilizes the abstract mind and creative imagination to imagine the future.

Normal events and occurrences stimulate the concrete mind, and extraordinary concepts stimulate the abstract mind through imagination. Here's another example:

Example 3: Visualization versus imagination

1. Visualize that you are relaxing in bed and reading a book. This is so obvious and normal that the concrete mind visualizes it easily.
2. Imagine that you are a point of light relaxing inside a tetrahedron. This process involves the abstract mind.

A person cannot visualize a tetrahedron if he or she has never seen one, and secondly, most people have never experienced imagining their bodies reduced to a small point of light.

When the concrete mind is too rigid as with some adults, they can hardly imagine unknown things. Children are more imaginative because their concrete mind is still flexible, so if you tell a child to imagine becoming a point of light, they can do it easily. But ask an adult to do it, and many won't even dare to close their eyes and cooperate to at least see if it is possible. The concrete mind immediately blocks the possibilities without even attempting to use the creative imagination.

An imaginative person is more creative, more futuristic and has greater potential to materialize bigger realities. Let's analyze this powerful formula:

> When a person has poor imagination and is very concrete-minded, they usually limit their mind to thinking of smaller possibilities, thus they end up with lesser probabilities resulting in a smaller reality.

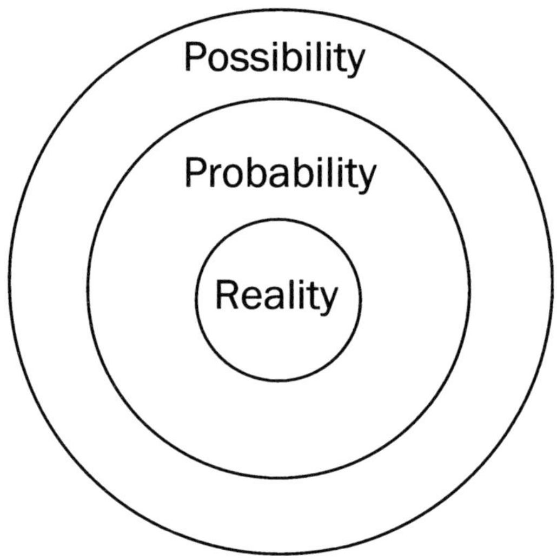

Whereas a person whose abstract mind is more developed has a better faculty of imagination and generally entertains bigger possibilities, resulting in a better probability and ending up in greater realities.

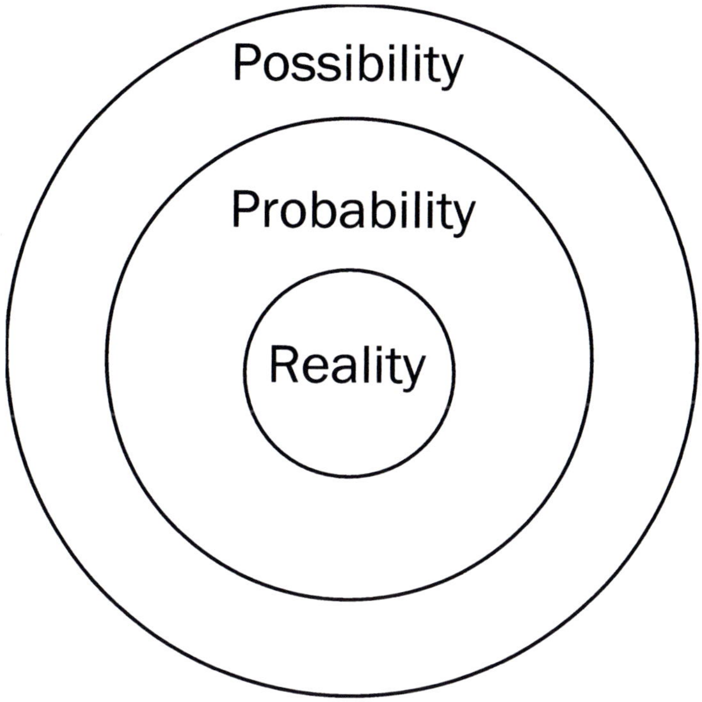

Advanced abstract thinkers are usually more forward-thinking and are dynamically positive because their mind allows their imagination to see ahead to greater possibilities that have never been explored before or are even opposite to common sense. This is one of the big contributions of the film industry, always showing humanity what lies ahead. Before cellular phones and electronic watches were made public, James Bond films were the advance promoters. Before Neil Armstrong and other astronauts landed on the moon, the movies portrayed stories of outer space exploration. Movies give examples of creativity and the power of imagination. They show the impossibilities that shape human beings' possibilities. Once human minds can visualize a result through imagination, a person can eventually do what they imagine and visualize. The effect of the movie industry, therefore, is like a double-edged sword. It can cut in both positive and negative directions: evolution or involution. That is the power of creativity and imagination brought by the media the, arts and the film industry.

Why is the process of imagination important?
1. First, because it is a tool of creativity and allows one to conceive new ideas.
2. Second, because it develops the abstract mind to advance its powers to tap into the future. The person becomes a visionary.
3. Third, creative imagination precedes the development of the intuitional faculty. This explains why more artists and mystics develop the intuitive faculty faster than the concrete-minded scientists.
4. Fourth, creative imagination is the faculty that brings a sense of humor. Humor is a byproduct of creativity and a wild imagination.

When the scientist learns how to use their imagination, not only visualization, great discoveries will result in the unthinkable!

When 3 minds meet and cooperate –
The concrete mind of the scientists
The creative imagination of the artists
And the guiding principles of the philosopher –
The future world will be a superworld!

Therefore, entertain life's possibilities beyond what the concrete mind can fathom using your imagination. This will keep you moving to the future with greater visions and reality – to the world unknown to most people. Think like an artist. If you are an artist, create works of art never seen by humans before, and bring human consciousness forward to the future. Paint or write from the future, and your work will be the lighthouse of new frontiers.

People with mental creative powers are always ahead. They are the pioneers, and they lead humanity to new adventures and unthinkable possibilities. These Divine co-creators from the ranks of humanity are the pillars of every country's structure, and they are the directing compass towards our future.

The power of this mental creativity, like creative inventions or discoveries, is truly pioneering, but the new ideas are sustained and born out of the knowledge of the past. Inventors and pioneers utilize and incorporate the wisdom and experience of the previous generations as a building block. Even Einstein claimed that Edwin Hubble's discoveries of the expanding universe made him adjust some factors in his famous equation of relativity.

Computers and the Internet came about thanks to the previous invention of electricity, electrical components and communications technology. The invention of electricity was made possible because of the invention of metals. More refined and advanced types of metals and compounds were discovered that innovated electrical systems into electronics, and then the computer came about. Without the electrical plug or electrical battery or a power source, the computer is nothing as you may have experienced when your laptop computer battery dies. Even Microsoft would not shine without PC hardware.

All the so-called technology pioneers can't create or innovate without these earlier creative discoveries and inventions. Without the creative pioneers of metal production, even the international air, water and land travel we enjoy today would not be available to us. So every inventor or discoverer of new technologies and sciences should have a sense of gratitude and should maintain humility to the creators and innovators of science and technology in each past generation. What we call modern science and technology are the by-products of the power of the mind of many civilizations. Close your eyes and imagine for a few seconds how it would be if we remove all things that were produced for us by business, science and technology creators. Where would we be? Start with your body, your home, car, the city. Close your eyes and imagine if we did not have their innovations and inventions. What do you see? Could we handle the situation for a week?

Practical techniques for mental power & its applications

> The mind needs mental exercise to grow.

There are many strategies to apply your creativity and mental powers. Many great thinkers and leaders have already mastered some of these methods. These are for those who are seeking ready-to-use guidelines:

Practical techniques for mental power & its applications

1. **Determine your point of interest** where you already have a good mental grasp or substantial experience. Is it in the field of leadership, technology, social and human sciences, psychology, politics, arts and music, culture, physical sciences, business, diplomacy, education, esoteric sciences? Select one or two fields.
2. **Use an integration of emotional and mental faculties and instincts to search for new ideas**. What ideas? Ideas that solve the common pervading complex problems of humanity and the world. Start with what you know and what you have

observed in your family or neighborhood. What do people you interact with complain about – a product, a service, an issue that has not been resolved or for which no business or organization has a solution. Write all of these answers down, and compile and gather data and information about these needs lacking attention or a supply of solutions.

> Great creative ideas and pioneering inventions come to humanity not because we merely want them, but because we need them. They come first to people who can access them because of high aspirations to help serve humanity and add more meaning to the world.

3. **Nurture the idea of helping evolve or solve something every day**, maybe a missing niche or link in the chosen field in which you have great interest.
4. **Identify where you can innovate or create**. Once you have felt in your heart and conceived in your mind that there is an area of specialization or subject where you can deploy some expertise, you can be sure that things exist to be discovered, to innovate or creatively replace already existing experts. Even in the computer industry, though it seems there are so many experts already, why do new and even better companies constantly emerge and succeed?

   For example, IBM was once admired and feared as the dominant player in the computer industry. Then Microsoft came along to dominate the PC market with Bill Gates' ability to innovate exist functionality. But Steve Jobs, a master of hardware, software and creative graphic arts and founder of the Macintosh computer in the 1970s and the iMac in the late 90s, is a leader in this industry – not a follower, not an innovator but a creator – despite Apple's ups and downs. Then there's Larry Ellison, founder of Oracle and database king. More examples abound: young Yahoo billionaire founders, Jerry Yang and David Filo, who created a great Web search engine powerhouse and Scott McNealy of Sun Microsystems, who introduced the user-friendly Java open programming language. I'm sure the computer and Internet industries have not yet been saturated because there will always be one or more great creative-minded people who've got the power to innovate and create new avenues of products and services that will be even better than the existing ones. With the above examples, I endeavor to present the idea of endless possibilities, new niches and untapped potentials for most people. Select an area of expertise or interest, then find or discover the missing untapped potential idea/product/service that you can embark on to develop and mentor.

5. **Make this new idea part of your vision and mission in life**. Create an aspiration that you are serving or want to serve, not only yourself, family or organization, but dedicate and surrender this project for the service of humanity and the Earth. This process is important because it opens your heart and crown energy centers and the spiritual connection that brings the real inner power in you to make this happen, the Power of your Soul and Divine Self. This naturally brings inspiration for you to proceed and build momentum.

   The initial impetus has to have breakthrough force and strong inspiration because when you start to physicalize a project, a lot of demoralizing conditions usually arise to squelch your desire to continue. Especially with big ideas, you need a big boost of inner power to drive you forward. Look at the space shuttle: The maximum required thrust and fuel is during the initial stages of the flight. Once the momentum is set, your continued interest and aspiration will be your lubricant to go forward, allowing you to finish what you start.

6. **Observe the Power of Silence for some time until you are ready to act**. Do not divulge your ideas, vision and plans to other people until they are solidified, especially to people who are not positive, supportive or trusting of you, even friends and family members. When people criticize or negate your plans, it weakens them and slows you down. It also creates doubts in you that can weaken your interest in your new ventures. Even your thoughts can be read or sensed by your peers, associates or people closest to you. Beware of jealous people. People can steal your idea and ruin it prematurely. Of course, in an established team or department in a corporation, it's a little bit different. You need to feel your way to tell the right thing to the right person or leader at the right time.
7. **Create a balanced resource group and team**. If you want to be the original proponent and leader of this project or idea, then after you are sure and prepared to reveal your vision and plans, you can create a balanced resource group who can help you with more detailed and accurate data gathering and fact finding assistance. You don't have to divulge the whole thing until you are sure of your team members. Reveal just enough for them to work properly as a team. If you choose to have only a limited partner for this project, then just choose

the best person to work with. Proceed with caution, but with certainty of success.

8. **Proceed and go deeper into your mental work** and continue penetrating all avenues and possibilities that can make your chosen project or field a unique, extremely powerful and much needed breakthrough. Do not spread your mission or idea too thin or in too many directions. This will not bring you focus unless you already have a big organization supporting your cause. For every additional goal or strategy, there is a tremendous increase in requirements for resources and manpower. Go for minimum resources and maximum output in the shortest timeframe.
9. **Go deep and not too wide!** Much failure in initial schemes and new projects is that the initiator and their team are too ambitious to cover everything in order to control without selecting specializations. Every increase in the scope of a plan requires a tremendous amount of energy and personnel and exponentially increases the period to accomplish it. It is better to dig a hole 1 mile/1.61 kilometers deep and 3 feet/.9 meters in diameter than 1 mile/1.61 kilometers in diameter and an 1/8-inch/.3175 centimeters deep.

For a pioneering breakthrough, there should generally be a selected niche or line of specialization. How did great pioneers and creators become globally known experts? Most of them were a proponent of only one small idea, equation or element. For example:

a. Madame Curie became significantly important because of one discovery she worked on for most of her life: the discovery and unleashing of the power of radium and radioactivity. I'd like to add a complimentary principle here that we discussed earlier as the process of conceiving new ideas due to humanity's need. Her discovery of radium and radioactivity came about right before the war. Luck and the power of timing made her project and services particularly relevant to the war victims (soldiers) since x-ray machines could locate shrapnel in their bodies. Thus the need for her service and related inventions was brought about by their usefulness.

b. Einstein's reputation and summary of his pioneering work is equated with the $E=MC^2$ equation, which made him extremely famous. He did not produce hundreds of equations or many areas of expertise. He dug a deep hole a few feet in diameter.

c. Billy Graham perhaps possesses a lot of talents, but he is best known for his specialized mission as spiritual crusader and mass spiritual counselor – using TV and media to create a breakthrough religious ministry.

10. Know your brand and stick to it. What is your summarized slogan and the motto of your project or area of expertise? There should be a few powerful words summarizing your new work or creation. By putting the underlying purpose and meaning of your new mission or specialization into one succinct statement or a concise slogan, it increases the power of the vision and mission. This is the power and magical use of words. The slogan and motto of your new work or emblem are the reminders of your new purpose.

    When you design a logo for it, create a very important symbol that summarizes the project. I don't recommend the color black color dominating the logo because according to my energetic observations, it inhibits the power and enthusiasm of the onlooker. Once you put your logo and mission statement–slogan into final artwork, the project becomes even more empowered. I call this "baptism of the organization", after which it really becomes an energetic entity. The logo and slogan affect the destiny of the mission to a great extent, so choose them properly. When people see you and your logo, they are your brand and symbol and an energetic summary of your project or mission.

11. **Continue the intelligent hard work**. Do not lose sight of the original vision and mission. It is easy to be sidetracked along the way. Patient persistence is required of all pioneering work because you are the one digging and shoveling the soil and rocky ground for the first time to open a new road towards the future. Do not compromise your inner values and noble purpose whatever happens along the way...It is not just the end results that are important, but also how you manage the process to achieve them. Earlier and faster results sacrificing inner values and compromising virtues are worse than a slower but spiritually clean achievement.

In almost every pioneering work, there is a degree of resistance from the populace or environment and from competitors.

Patient persistence and courage are needed for pioneering work: The Rachel Carson Story

Let's look at what happened to a pioneering American environmentalist, Rachael Carson, as related by Peter Matthiessen.[15] Rachel Carson wrote the famous book, *Silent Spring*, in 1962, which was a compilation of her scientific research, investigation and findings on the negative and poisonous effects of chemical pesticides on the natural environment. Her patient persistence and clear vision of her purpose to help save the natural environment bravely pushed her against the violent intimidation and threats of lawsuits by big corporations. Also, before she became a well-known author, her literary works were said to have been rejected by 15 magazines, including *Saturday Evening Post* and *National Geographic*. So we can see how much of a persistent, single-handed effort and force she employed to finish the goals set before her. Later her works were published and heralded as a work of environmental service. If there are obstacles that challenge your creative path to unleash new services and breakthroughs, just remember this woman born in 1907 who was a brave environmentalists before environmentalism as a movement existed.

12. **Power of Sacrifice**: If you think the end result and influence on evolution of your newly chosen creative mission is greater than your life, it is worth a sacrifice of most things in your life in order to leave a legacy. This principle was always part of guiding light for the lives of great leaders, great Saints, Sages and World Servers. What do we usually sacrifice? Lesser things for bigger goals: life, time, effort and our old personal preferences. The power of our achievement is directly proportional to the levels of our sacrifice in life.

These above mentioned strategies to apply the powers of the mind for the creation of new and important, if not extraordinary, projects and services will allow us to become extremely useful in society. These techniques are presented to also address the need for mystical groups and untrained individuals on the spiritual path to commence a more practical use of the mind to excel in their chosen path or careers, to flower where they are planted.

Businesspeople, scientists, educators, politicians and more mentally developed people will also benefit from this section because it offers some potentially useful hints and strategies to help already great achievers perform even better and more efficiently. The section on the development of abstract mental powers can be a good study and periodic experiment to enhance and align what you already know.

How the mind works: The mental power of focus & awareness

This is one of the last faculties of the mind to be mastered because it involves the development of the concrete mind, abstract mind and augmented by the sensitivity of the heart.

In our previous discussion, we have stated that there are two energy centers that activate and regulate the mental faculties:

1. Throat center – concrete thinking faculty
2. Ajna or mid-brow center – abstract thinking faculty

The mental body has energy centers corresponding to the vitality and emotional energy bodies. The mental body is made up of thought substance and vibrates at a higher frequency and also integrates and extends beyond the vitality and emotional bodies. By activating the throat and ajna energy centers related to mental functions, their psychological powers are awakened further. To make this more effective, it is advisable to first purify the mental substances of negative thoughts and intentions or thought patterns obstructing the mind from growing. After that, we will activate and revitalize the throat and ajna centers through breathing techniques. To integrate and balance the concrete and abstract mind, we will simultaneously focus on them and do the inner breathing. The heart center will also be stimulated to sensitize the emotional intelligence and compassion to use the mental powers only for good purposes and thus a new faculty and inner power will evolve into Love-Intellect. This is required to develop the power of focus and awareness.

The mental power of focus and awareness

First procedure: Purification of the emotional & mental bodies

(7 minutes, twice weekly)

1. Sit in a comfortable position with your spine vertical and your feet flat on the floor. Close your eyes to enhance the concentration.
2. Visualize that you are in front of an ocean and breathe slowly and deeply. Relax. (30 seconds)
3. While breathing slowly and deeply, mentally intend to breathe out and expel all negative stress, tension and feelings to the ocean. Imagine that the ocean is absorbing them. Just relax. (2 minutes)
4. Continue the breathing technique and mentally intend to breathe out and expel all negative thoughts, intentions, confusion, mental disturbances and any harmful thoughts against yourself or others. Imagine that the ocean is absorbing all of these. (2 minutes)

Second procedure: Revitalization and activation of mental & emotional powers

1. Concentrate on your heart center and breathe slowly and deeply (7 times).
2. Concentrate on your throat center and breathe slowly and deeply (7 times).

3. Concentrate on your ajna center in between your eyebrows and breathe slowly and deeply (7 times). Simultaneously concentrate at the ajna, throat and heart centers and breathe slowly and deeply. (7 times)
4. Mentally do the following affirmation 3 times:
   I am a creative and intelligent person. I will use my creativity and intelligence lovingly and properly.
5. Practice this technique after a shower or bath and before eating full meals. It may not necessarily be safe for smokers and people with heart conditions to practice this technique. If you are not certain, consult your medical doctor before you start.

   Note: The MDP Synchronized Breathing Technique (5 minutes) can be done before this technique to achieve greater inner purification results.

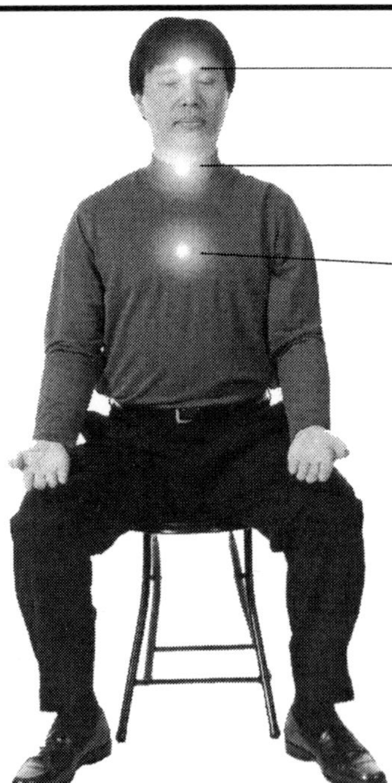

Method to develop the mental power of focus and awareness

Breathing technique to develop love-intellect faculty

Figure 22

When this emotional-mental power development technique is done regularly twice a week, there will be a notable change in your thinking capabilities, with more integrated concrete and abstract mental faculties and emotional intelligence. This quality is the next goal for human development: the good qualities of an Albert Einstein, Bill Gates and Mother Teresa all in one personality. How about that?

There are three groups of thinking styles:

1. People who are too focused, meticulous and like details
2. People who are too general and like principles and philosophies
3. More advanced thinkers who are both naturally focused and aware, an integration of the above two natures

Detailed thinkers

The first group, the detailed and concrete thinkers, tend to have an active and big throat chakra and more left brain activity. They are very good at concentration and concretizing projects, and you can rely on them with specifics. They tend to be more accurate and scientific. Day-to-day affairs is the name of their game. The problem is that they are lost in details and sometimes forget the general plan and long-term goals. They tend to be telescopic in perception and to scrutinize your faults through a microscope. The seed of failure in this group is the emotional body, which is not necessarily developed or has not been used regularly and thus has atrophied. Usually, they are relatively cold and indifferent and are introverts or enjoy working alone. These people are good implementers. They usually finish what they start. They tend to be subordinates or researchers, or they are not the center of the business. If you are not this kind of person and your boss is an extreme concrete type, you are in trouble. Artists and mystics are the opposite of these personalities. Both groups can learn from each other's style through osmosis or interaction.

The philosopher and artist types

The second group is the philosophers and artist types. They are good at generalities, principles and creative abstractions. They are not as pragmatic as their detail-oriented counterparts. They talk of ideas, but a lot of them talk about ideas for 20 years and never act on them. These people are more abstract thinkers, and the ajna center is more active than the throat center. That is why you can encounter both types in a business meeting: The results-oriented people bring a thick file of data and graphs into the meeting, talk immediately about results and ask the bottom line and 2) The abstract thinkers bring a thinner file and maybe a book with quotations marked to remind you of the principles of success. These people are also always aware of the vision statement and general long-term plans. The more abstract-minded group usually checks and balances the quality of processes in an organization. They see longer-term implications of the company's role in the future. They are practitioners of awareness and sensitivity and are the visionaries.

Integrated thinkers

A third type of thinker is a person who integrates focus and awareness.

> When the concrete mind, abstract mind and the heart are active and integrated in a person, you have the best of the best leaders or thinkers around you.

Integrated people are motivators, inspirers and transformers, yet have very good acumen in managing personnel, resources and investments. They are more advanced Souls whose long ancient experience has endowed them with Love-Intellect integration. Most of these great thinkers specialize not only in the what's and how's of things, but also the why's. The incoming generations of top thinkers will express this integrated faculty. When the educational systems catch up on the emphasis of inner practical values, plus the training of the abstract mind in the curriculum, lessening and simplifying the learning process of the concrete mind, then the mass production of real genius – heart-mind thinkers and leaders – will be common everywhere.

The mind is an excellent tool, and it's a great idea to use the mind regularly and to enjoy sharpening it. It is better equipment than any technology to organize our daily life continuously, harvest the lessons of the past, accurately predict the future, make good decisions, solve problems properly, innovate our world, create a new culture…you can name unlimited applications.

Scientists have made maximum use of their concrete minds, the philosophers their abstract minds. And your mental powers have evolved just by reading this section of the book about the mind. Isn't it said that knowledge is power? Therefore, by increasing the amount of positive information in your mind, you are no longer the same. This is accurately portrayed by Oliver Wendell Holmes in his famous statement:

> One's mind, once stretched by a new idea,
> never regains its original dimension.
> - Oliver Wendell Holmes, American author, poet & physician

Your continuous learning is always the sign that you are alive and evolving, and you can be sure that with patient persistence, you will grow towards the light that illuminates your whole being, allows you to be the guiding light for others. And you will become the light for the dark corners of the world. It's just a matter of time before the mind can be mastered. It has already begun. But the pursuit of all the needed knowledge and learning takes a lifetime. According to a Chinese proverb:

> Tell me, and I will forget.
> Show me, and I will remember.
> Involve me, and I will understand.
> - Confucius, Chinese philosopher

Therefore, I've shown you how the mind works, the ramifications, elements and applications of mental powers and how to develop them rapidly and properly. Apply and practice the techniques so that you will "understand" through your own personal experience.

Section D Psychic Powers: Principles & Techniques

I. Psychic power experiments ..... page 182
- A. Experiment 1: Refrigerator experiment ..... page 182
- B. Experiment 2: Office or house experiment ..... page 183
- C. Experiment 3: X-ray vision of hands & fingers ..... page 184

II. Attitudes for psychic power development ..... page 185

III. Different types of psychic powers ..... page 194
- A. Psychic powers to be developed by humanity ..... page 194

IV. Basic requirements for psychic power development ..... page 195
- A. Energy level ..... page 195
- B. Quality of energy ..... page 195
- C. The condition of the energy screens or sieves ..... page 198
  1. Functions of the energy sieves ..... page 199
  2. Different types of psychics ..... page 200
- D. The condition & development of the energy centers ..... page 204
  1. Chakras for psychic powers ..... page 205

V. Psychic power development ..... page 208
- A. Clairsentience & psychometry ..... page 208
  1. Procedure 1: Internal martial arts hand sensitizer ..... page 211
  2. Procedure 2: Applications of clairsentience through Hands evaluating etheric energy ..... page 215
  3. Procedure 3: Comparatively sampling the vitality aura for light (higher) or heavy (lower) vibrations & high & low energy ..... page 216
  4. Energy level detection practice ..... page 217
  5. Practice to feel & detect emotional & mental substance and their qualities ..... page 219
  6. Sensing other people's qualities through auric sensitivity & instincts ..... page 220
- B. Etheric vision ..... page 222
  1. Experiment 1: Seeing the aura of your fingers ..... page 222
  2. Experiment 2: Seeing the aura of the head ..... page 224

3. Experiment 3: Seeing the aura of sunset...................... page 225
4. Experiment 4: Etheric x-ray vision................................. page 226

C. Clairvoyance... page 230
1. Experiment 1: Candle flame practice........................... page 232
2. Experiment 2: Advanced method to see the auras of the sun at sunset... page 235
3. Experiment 3: Seeing the energy bodies and energy centers of human beings........................... page 235
4. Experiment 4: Seeing the auras of animals, plants & natural crystals... page 238
5. Experiment 5: Clairvoyant practice with distant subjects... page 239

D. Clairaudience.. page 240
1. Method to enhance clairaudience or inner hearing..... page 243

E. Telepathy... page 245
1. Practice 1: Conditioning your telepathic equipment & developing your concrete mental telepathy...page 250
2. Practice 2: Conditioning your telepathic equipment & developing abstract mental telepathy... page 252

F. Precognition faculty.. page 257
1. Precognitive psychic faculty development technique.. page 260

G. Channeling, automatic writing, drawing or movement...page 262
1. Practice on channeling.. page 266

H. Telekinesis or psychokinesis.. page 267
1. Experiment 1: Energy telekinesis.................................page 268
2. Telekinesis technique to bring love, harmony & peace... page 271
3. Experiment 2: Physical telekinesis by pushing a person through his/her aura........................ page 272

I. Emotional & mental body projection...................................page 274
1. Maximizing inner experiences during sleep technique...page 280

J. Teleportation.. page 284

Section D
Psychic Powers: Principles & Techniques

When we talk about psychic powers, we are referring to faculties that are natural but latent in the vitality, emotional and mental levels of human beings and not apparent to the common sense of humanity as of the writing of this book. But today many mystics and psychics in different countries are exerting their presence and talents to be recognized and felt. In fact, their careers as mediums and psychic readers are starting to be accepted, even in Western culture. In the East, psychic power unfoldment is generally within the accepted tradition. In the West, there is an increasing number of children who have already partially developed some psychic powers, but our modern society still labels these natural faculties as sixth sense or extraordinary powers. Movies like "Sixth Sense", "Unbreakable" and "Lord of the Rings" have been awakening the public to the existence of these psychic faculties. As we said earlier, the film industry play have a big part in forecasting the future potential of humanity in which – within this century – psychic power development will probably be part of education and religion.

Let's experiment with simple experiences and proccsses before explaining the regimen and principles of psychic faculties?

Psychic power experiments

Experiment #1: Refrigerator

1. After reading these instructions, sit in a comfortable position and close your eyes. Breathe slowly and deeply several times and relax.

2. Use your creative imagination or visualization. These 2 processes are the initial training for the proper awakening of the psychic faculties. You should always be in a relaxed state. In psychic development, do not use concentration or will. You need to use feelings and sensitivity more, relax and have a more surrendering attitude. Do not push or control results or fabricate answers.

3. Just allow things to happen: Imagine that you are in your kitchen. Just let your consciousness go there with very gentle slight intentions. Without opening the refrigerator's door, look at the contents inside as if your eyes have x-ray power. See what is inside. Be aware that you are seeing it with infrared technology or see through it with a high vibration consciousness; you can infiltrate and penetrate things as if the refrigerator door were transparent. Look through the refrigerator. Know that it is easy and doable without forcing the process with your mind. You are soft and gentle and light in your feelings. Take your time... What do you see?

4. With slight intention, imagine that your hands are inside the refrigerator and that your hands are like a form of

light. They can sense anything inside the refrigerator. Feel it and let the data or profiles gently appear in your imagination....

5. Imagine that you are putting your nose inside the refrigerator. Smell anything inside. How does it smell? Relax and just allow data and profiles to come in to you...

6. Imagine that you are biting into some food from the refrigerator. Take your time...How does it taste? Register in your consciousness whatever taste comes into your mouth. The secret is to allow the information to come. Relax. Wait. Be patient.

7. Imagine that you are opening the refrigerator door. Then slam it shut. Do this twice. How does it sound? Relax and register the experience.

What was your experience? Did you have inner correspondences to your outer senses? Let's do a second experiment.

Experiment #2: Office or house

1. Imagine that your consciousness is inside your bedroom. What do you see? Let the profiles and data come in without fabricating them...Relax and allow the gentle appearance of the subjective forms. Register the information.

2. Now, project your consciousness to your office or any familiar location...What do you see? What appears in your imagination (not visualization)? Who is in there, and what

are they doing? Again, do not fabricate but allow the gentle appearance of profiles or data...Relax...Just patiently observe and then things will register.

Note: If you are in your office, then try transporting your consciousness to your home or somewhere else. Follow the same guidelines.

What was your experience? Did you see any movement of people in your mind's eye? Did it appear as a film or movie?

Experiment #3: X-ray vision of hands and fingers

1. Now, let's look at the energy of your hands and fingers. Keeping your eyes closed, be aware of your inner vision...Allow your imagination and consciousness to penetrate through the fingers: first through the skin, then the muscles and their blood vessels, then the bones – all of them like transparent tissues...Relax, breath and allow the appearance of profiles and data in your consciousness. Be patient...Just wait.

2. Continue with the energetic profiling of the hands and fingers. Light interpenetrates and surrounds the physical hands and fingers. Allow the gentle appearance of light and colors or auric profiles. Do not fabricate or visualize...Just let data and profile appear. Take your time and relax.

Well, what do you think? How do you feel? What are the inner experiences? When you are answering these questions, your mind may doubt or reject the input or it may seem like you know, but cannot label or express the experience in words. These reactions are normal for beginners or even untrained psychics. You should use the following attitudes to train your psychic powers easily and properly:

Attitudes for psychic power development

1. **Be neutral.** There is no need to be in a state of belief or disbelief. When you are using the process of reason, your concrete mind is in duality, therefore the ability to see subjectively can be difficult.

2. **Do not use any will, pushy intention or concentration.** Use sensitivity of feelings and gentle awareness more.

3. **Relax and maintain gentle slow breathing**. Be in a calm but energized mode.

4. **Use imagination more than visualization**. Visualization uses concrete mind power more and usually fabricates or constructs thoughts or visuals. It partially uses mental will, which runs counter to psychic power. Imagination allows the appearance of a parade of visuals or energetic forms. It is similar to daydreaming and is connected to gentle awareness and feelings. Visualization only sees things known before. Imagination can see things or events that have not yet been experienced or seen.

5. **Wait and be patient for the appearance of profiles or the developing data**. It can take time for the mental or emotional aspects of the brain to interpolate and interpret data, especially if the brain and the mind have always been used to think, evaluate and process data logically. Very objective, mental and willful people need to relax and be more gentle because their training has made them very physical and mental. Psychic powers use inner perception and are more subjective.

6. **Be aware that you exist in an interpenetrating vibration or energy that is made up of invisible light**, and everyone, animate or inanimate, is floating and immersed in it. This awareness and new program in your consciousness is important to dispel the illusion of just plain material existence. That's why when people are dreaming they still think that they have to open the door of a building to enter it. Since they are in their dream body - emotional and mental bodies - they are more refined than the door, thus they can pass through without opening it. Why do people have this conception? Because most people don't even have the idea that we have invisible bodies, which are made up of very refined energy and light vibrations, that we are not just made of physical, liquid and gaseous matter. Once this concept is part of the person's consciousness, the inner world or the subjective realms are easily revealed and experienced. Now, if I use the word consciousness, it refers not just to mental consciousness, but also includes the emotional intelligence. In psychic faculty development, the registry of experiences is not always through the brain or mind. That is the common blockage in starting practitioners. It is easier to start with feelings and energetic sensitivity than mental awareness. This explains why most emotional people are energetically sensitive and are more easily trained to develop psychic powers and inner sight than mental type people.

Now let's evaluate the process and results of the experiments:

1. **Experiment # 1: Refrigerator**

   The five senses are usually the receptive mechanism of the objective world registered by your auras, emotions and mental bodies as concrete data – which is usually processed by the mind. I used the word usually because in more advanced people, the registry of information on chemical, hormonal and electrical signals of the body is overtaken or bypassed by the information or signals of the vitality body and its energy centers, *nadis* and meridians. *Nadis* are the energetic equivalent of the nerve endings. Meridians are the pathways of energy connecting the chakras and the nadis.

Therefore, what is familiar to most people and medical scientists are the physical, chemical and electrical processes of the body, but energy healers, more advanced meditators and psychics are familiar with the energy or vitality body, other auras and energy centers. Internal martial artists start sensing this energy, which they call *chi* or *ki* or *prana* in yoga.

As you may have noticed, there are also inner correspondences of your five external senses. Most of you might have imagined and even subjectively looked through the refrigerator, saw what was inside, imagined the colors of the contents (inner seeing), smelled the food inside (inner smell), subjectively touched the food (inner sense), subjectively tasted the food (inner taste) and subjectively heard the sound of the refrigerator door (inner hearing). Even if you think it's just your imagination, that's a great start. Even with closed eyes, you imagined, sensed, smelled, tasted something – the initial step in developing the inner faculties of your etheric, emotional, and mental bodies – called psychic powers. Some people might even see the aura and energy of the contents of the refrigerator immediately. Great! For those who have not sensed or experienced enough, this experiment is a great start. Don't worry – experiences will unfold.

2. Experiment # 2: Office or house

This second experiment is a little bit different because it enables you to transport your consciousness to another place. It can be done as a projection of your inner sight like a camcorder or telescope, or it can be an out of body experience, like when you sleep or meditate. Either of them are fine and acceptable methods.

This experiment will enable you to experience registry and appearance of movements, including people's dynamic activity. If you think you're just visualizing, you're not! What you registered about certain people or sequences of activity were not really a product of visualization; they were more a product of imagination, similar to recordings by a video camera. This experiment allows you a continuous array of inner movement called your imagination. Visualization is a more fixed, non-moving and non-continuous visual. The second experiment gives you a process to ponder on.

3. Experiment # 3: X-ray vision of your hands and fingers

Since you started using x-ray vision through the refrigerator in experiment #1, it should have been easier to follow the concept and process of seeing though your hands and fingers when you did the third experiment. If you experienced seeing blackness or darkness, it's all right; that's a start. If you start seeing a little profile of light, that's fine. Or if you start seeing more shapes, forms, light or energetic movements, it is also a great start. This experiment is important to de-program the old belief and memory of the mind and brain that your fingers are just solid physical forms. By awakening the principle of invisible energy and light, plus the existence of an interpenetrating vitality vibration enveloping the hands and fingers, there is already a breakthrough in your psychic consciousness: The new program from these experiences superceded the limited concept of physical reality.

Learning to develop your psychic powers is easier through practical applications and doing experiments rather than just talking about its principles, but for the experiences and experiments to be understood and appreciated, let's elaborate more on the processes and techniques required to unfold the psychic powers.

Developing psychic powers

First of all, why do we need to develop psychic faculties and psychic powers?

Why do I need to elaborate on this concept? Well, as I've said earlier, we already have these powers latent in us, but in most people they are dormant, and if some have them, they are not completely manifested to their highest potential. But why develop psychic faculties and powers? The reasons are numerous:

- **Development of extraordinary aptitude for:**
  - Perceiving truth and higher realities
  - Good precise instincts of knowing
  - Perception of higher insight beyond the mind and common sense
  - Improving visionary or precognitive faculties to sense or predict incoming needs
  - Sensitivity to the energy bodies to improve awareness of subjective realities and worlds
  - Improved decision-making and problem-solving skills
  - Channeling higher ideas for personal growth and service

- **Enlightenment of our psychic potential:**

  Most psychics are born with psychic faculties, but there are some problems inherent in their untrained use.

  - Some psychics may misuse psychic faculties because of ignorance about their proper functions.

- They may mislead people because of wrong interpretations due to what they read psychically about their subject
- Some are contaminated by negative entities or foolish spirits for not invoking the right spiritual guide, or they are overshadowed by negative beings without even knowing that a negative force is using them.
- Many children have been born with some psychic talents and they unfold them early in childhood, but usually these faculties are improperly shut off by scared parents, teachers or religious counselors. So whether you are a born psychic or not or have a psychic child, the scientific training, proper understanding and unfoldment of the faculties and powers is to your advantage. It is ignorance that can bring their misuse or destructive applications. Also, when adults or children can be trained and guided to use the psychic faculties, they have many important applications to improve the quality of life, accelerate the learning process and maximize performance in personal growth.

The new generation of modern children, who are the early forerunners of psychic experience, will be the future adults who will witness and validate the inner realms and the world of invisible energies. With their help and outstanding experiences, educators and scientists will be induced to investigate and establish research on the psychic phenomena that will eventually unfold as part of the new educational system.

By initiating the possibilities and concept of psychic powers to the scientists, businesspeople, educators, politicians, social reformers, artists and religious groups, some of these people will recognize that many of their children's experiences and talents are normal. Psychic powers and faculties will be accepted as a natural aspect of evolution and will become common.

When many top leaders and celebrities understand the inner workings of subjective realities and universal principles, then a more enlightened approach to humanity and world problems will be unfolded with greater understanding and better long-term solutions. For example, most private, public and governmental organizations are still focusing only on physical environmental ecology and sanitation while the bigger problem is that the pollution and contamination of the Earth and of most people are at the emotional, mental and vitality levels. People are swimming in an ocean of unpleasant and negatively programmed mental energy and emotional Earth vibrations accumulated through eons of time. The vitality of the Earth, the ground, air and water energies is also affected not only by chemical and gaseous contaminants and pollutants, but also by invisible substances and frequencies.

When humanity understands this, the next step is to clean up the subjective vitality, emotional and mental worlds in order that:

- The minds of humanity might think more clearly and positively
- The emotions of people be purified of fear, anger, stress, hatred, jealousy, greed, selfishness and violence, which are all just negatively programmed energy substances and entities
- The stale contaminants and diseased energies of the auras of humanity, animals, plants, minerals and elemental kingdoms be cleansed, revitalized and transformed into healthier energies and vibrations

In the future, with the awakening of psychic energy awareness, the energetic and psychic ecology of both the visible and invisible worlds will become a priority subject of research and concern. Until the mental and emotional energy fields of the Earth are properly cleaned or "ecologized", the negative programs created by the dark ages of humanity are pollutants and poisons to its inhabitants. Humanity is like a fish swimming and residing in the ocean of inner vibrations. Thus, forests are destroyed because some people are polluted with greed, and many people litter because of their ignorant and unenlightened minds. Both microcosmic and macrocosmic cleaning, revitalizing and transformation involving humanity and the Earth are much needed – not in the future, but now, immediately at all levels. But if we don't do it, I'm sure that our more enlightened children and future generations will do it because of they will realize it through their advanced psychic powers.

With the proper development of psychic faculties and powers, education will change its curriculum and way of teaching. It will become much quicker to learn and comprehend the different subject matter and fields of study. Blockages of students in their learning experience will be diagnosed and rectified easily. Learning the objective and subjective realities will be integrated. Therefore, the assimilation of experiences, through both knowledge and wisdom, will be faster. Linear education and experiences will be obsolete.

The nature of the energy system, the Soul and the Spirit will not be a matter of phenomenon or belief, but common sense to most students. Therefore, spiritual and inner development will be technical, scientific and pragmatic in approach. As a result, incoming civilization will be psychic and intuitive, and also scientifically minded and materially grounded.

Business, as well as politics, will be based on integrity, trust, and honesty because future generations will easily sense and psychically see negative and unworthy leaders and business partners. Inner values will be real and common qualities, even in business and politics. People will elect and support advanced Souls and leaders because they will be seen to have real integrity through their developed chakras and energy centers.

Health care will be totally different. It will be an integration of allopathic medicine with energy medicine and healing, and much progress and investment will be focused on preventative health care rather than on corrective, palliative measures and pharmacology. I even envision that genes and DNA will be altered with more advanced healing methods affecting the past and the future.

> The development of psychic powers will bring in a new way of dealing with health. The blind will be taught how to read psychically and to see and explore the world even with their physical blindness. The deaf will be able to hear in different worlds and enjoy different "inner world radio programs". They will also use telepathy to hear messages and to communicate with others.

Instead of children consuming many hours watching TV, they will spend a fraction of their time exploring the realities of the inner world, learning in the inner realm. Their psychic explorations can allow them to learn in a much shorter period many of the experiences that normally take many years to study. They will have more dircct, "inner world Internet access", without physical modems and computers. Technology will focus on developing different products that address their clients' totally different needs.

Physical science and education will be working closely with esoteric sciences and will eventually become an integrated science. Then the true secrets of creation and evolution will be understood clearly with or without gadgets.

The exploration and explanation of so-called space and extra-planetary realms will be connected to out-of-body experience – seeing the glory and grandeur of spiritual beings – and external physical travels will be matched or even superseded by internal travel to explore and tour the inner world.

I'm presenting more possibilities than we are likely to accomplish in just a few decades during this period that marks the beginning of humanity's golden age, but we have to begin from the basics – to first understand the invisible realities and psychic faculties and how to develop them faster for better service and maximum unfoldment of our individual and group purpose. Through the refinement of the psychic faculties, we will also be able to improve our discernment, decision-making and problem-solving aptitudes. This will result in the eventual reduction of mistakes that create recycling of more suffering and negative karma, which delay our evolution as a whole.

Psychic Faculties to be Developed by Humanity	
1. Clairsentience and psychometry	Feeling and sensing energies and vibrations of objects
2. Etheric vision	Seeing the etheric energies and energy centers and, at an advanced stage, seeing with physical x-ray vision (with eyes open)
3. Clairvoyance	Seeing etheric, emotional and mental bodies or vibrations
4. Clairaudience	Ability to hear energies and sounds of the etheric, emotional or mental substances and vibrations
5. Telepathy	The ability to project, receive and read thoughts and ideas, especially at the mental energy levels
6. Precognition	The faculty of seeing or knowing future events before they happen
7. Automatic writing,	Bypassing the mind and emotions by bringing

drawing and channeling	information through the physical and etheric bodies
8. Telekinesis or psychokinesis	Ability to move energy and objects
9. Emotional and mental travels	Out of body travel or projection
10. Teleportation	Ability to transfer oneself from one place to another location or level

Basic requirements for psychic power development

Certain factors are important to know and understand before developing psychic faculties and powers:

First requirement: Energy level

Like video recorders, x-rays, ultrasound, EKG equipment and sensing devices, all psychic and spiritual faculties require an electrical power supply to work. The same is true with psychic reading and perception. These processes require a substantial amount and level of energy for them to work with more accurate perception. The higher your energy level, the easier it is to produce accurate results.

Second requirement: Quality of energy

To be able to switch on the psychic faculties and powers is just the first step. The need to interpret or accurately perceive what you sense is another level. Therefore, the substances, vibrations and energy centers of the different energy bodies – etheric, emotional, and mental – should first be purified and decontaminated of negative energies created by yourself or transmitted by other people, beings or the environment. Consider that one of the functions of the different bodies and your energy centers is like a vibrational and information, photography and trapping mechanism – and these energy bodies and chakras have intelligences that absorb, assimilate, process and transmit the data to your consciousness. Some mediums and psychics in a trance state can collect information without using the mind and brain as the interpreter. In fact, some of them do not know what transpires in the psychic reading or the messages they deliver. Sometimes, the person has never before learned the topics or subject being channeling. Very advanced human beings can bring new information at will without previous study or data on the topic. This also occurs with some high quality automatic writing, channeling or drawing.

The quality of picture or captured information through a video camera is as good as the clarity of the lens. If the lens is dirty or cloudy with dust, the captured pictures are contaminated by impurity in the lens. The same is true in perceiving through the energy bodies and the energy centers. They are also subjected to many contaminants and negative influences. These impurities can be:

- negative thoughts, illusions, and entities in the mental body and its centers
- negative vibrations, feelings, and desires in the emotional body and its centers
- negative energies and entities in the vitality body and its centers
- negative vibrations and entities in the transitioning zones and protective layers between the different bodies serving as energy screens or sieves. Esoteric scientists and healers sometimes call them protective or atomic webs. This will be discussed later in more detail in Section E on Spiritual Powers.
- negative energy or lower vibration of the immediate environment and people around the psychic while reading

If a room is filled with smoke and low-level light – even if your video lens is clear – the pictures you capture will be colored or affected by the environing conditions. The same thing occurs when seeing or sensing with your psychic faculties in an energetically negative environment, especially with lower frequency people around you. In fact, in most cases, the person being read and the surrounding people's thoughts, emotions and intentions affect the reception and interpretation. The psychic reader is often reading the thoughts or wishes of the person being read and does not even know that this is so – and, of course, the subject is happy to receive information from the reader that parallels what they want to hear. This is where mediums commit many mistakes and disservices.

The inner subjective world is composed of negative as well as positive beings, entities, elementals, vibrations, and forces that come from within and without the planet. Many types of existences are present in a world within worlds, in dimensions interpenetrating other dimensions. Humanity's energy bodies and chakras have been designed and equipped with screens or sieves that do not allow admittance of unsuitable vibrations into the different bodies. The problem in many cases, like drug addicts, alcoholics and insane people, is that these "energy screens" are partially damaged either at the energy center or in different parts of the auras, like window or door insect screens with holes. In some people it is worse because the majority of the energy screens in their auras are damaged, which leaves them with no defense against unsuitable energies.

There are many causes and reasons that "energy screens" become damaged, punctured or energetically torn: excessive use of tobacco, alcohol, recreation drugs, very toxic elements, long exposure to severe negative emotions and psychic attacks or black magic.

Some advanced spiritual students and initiates use a method of burning or gradually removing these energy screens to continuously integrate the different energy bodies, which produces superior inner powers and faculties. But this is only safe and possible when the person is very advanced and the personality has done its strenuous purification of all the energy bodies and the contamination from the ancient past. It is also safe if a qualified teacher guides and closely monitors the student. Without technical guidance and spiritual protection, this is really what we call "playing with fire".

In some older schools of mysticism, the teachers and students use what they call "power plants" and natural compounds as an agent to burn the energy screens. These are hallucinogenic herbs or potions that are similar to recreational drugs. I cannot recommend this because most people live in dirty, stressed cities or environments, both psychically and physically. If practitioners living in cities use this artificial method to open up their psychic faculties, they can absorb negativity from the environment in their auras and bodies, resulting in either insanity or spiritual possessions. Once these energy screens are damaged, it internally weakens and drains unprepared people, so they are in great danger.

Not too many healers or therapists can handle these cases yet. Most of them do not even know the esoteric components of the human energy system. One group that can handle damaged energy screens is the Pranic Healing group of one of my early spiritual mentors, MCKS. My team of consultant-healers also handles difficult cases.

Third requirement: The condition of the energy screens or sieves

We have studied and identified the different energy bodies and their centers – which are made up of varying intelligences, intensities of light, frequencies and substances. And we have seen that all of these energy bodies and centers have separate functions and connect to different energy frequency stations or dimensions on the Earth's energy field as follows:

1. The human vitality or etheric energy body is interpenetrated by the Earth's etheric aura and functions using this substance and vibration.

2. The human emotional energy body is enclosed by the Earth's emotional field and aura and functions within this emotional world.
3. The human mental energy body is interpenetrated by the Earth's mental energy field and aura and functions within this mental world.

Functions of the energy sieves

The energy sieves, serving like thresholds that separate the different bodies, have several functions:

- They are intelligent energy sieves that screen incoming energies, allow harmonizing vibrations and reject unwholesome energies or negative entities and forces.
- They also act as separating veils that do not allow your psychic faculty to always see or sense the other side of the veil – meaning they are like shutters on your camera or video recorder. They act as a curtain to protect the person's consciousness and development from what they are not prepared to see. The inner world, including ugly and negative or demonic entities, is not revealed to you psychically until you have a use for such sights or are spiritually mature enough to apply certain esoteric service or duty. Herein lies the different levels and maturity of psychics.

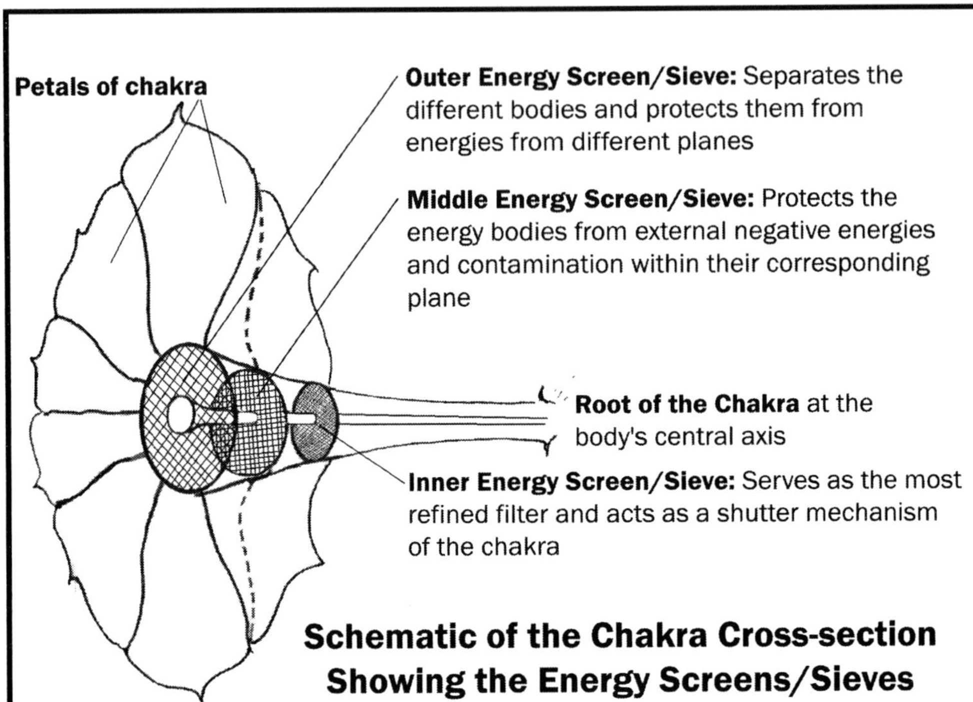

Schematic of the Chakra Cross-section Showing the Energy Screens/Sieves

Figure 23

Different types of psychics

First type: Unconscious psychics

This group is born with sensitivity and some partially unfolded feats, but they are unable to use them at will, or the psychic power comes and goes. They are sometimes unsure of how to interpret their psychic readings or are scared because they are not trained and do not understand what they perceive. If they are spiritual, they belong to the mystics. If they are undeveloped and unclean, they can be misused by negative beings or forces, and they normally abuse their faculties for selfish ends. Many unconscious psychics are either in business, education, science, arts or underworld groups. The involutionary forces use many of them to implant their negative plans – and some of them do not even know it.

In this coming era, more and more people – both those who are spiritually awake and others with energetically damaged open auras and centers – will incarnate, and their conditions allow them to have psychic abilities. The less developed unconscious psychics tend to be used by negative groups or forces for their own ends. This situation can lead to the misuse of inner powers, causing psychic wars and spying. This has already occurred in some periods of our modern history, but harm or misuse can be averted if the people with these talents, especially the incoming advanced children who have inborn powers already, are educated and guided in the true development and use of this power and faculty for service.

Ignorance of unfolding inner powers can lead to the misuse of powers, which is a cause of very expensive wars, destruction and chaos. It's like giving a shotgun to a child or a grenade to a teenager. Since unfolding the latent psychic power to be able to live consciously in the external and inner realms is part of human development, then we should not fear psychic power. Instead, we should prepare ourselves and the incoming generations for it. This is one of the purposes of this book, especially this section.

Second type: Energetically damaged psychics

People in this group tend to be victims of forced opening of the aura, centers, or "energy sieves". The causes and triggering factors are:

- excessive consumption of alcohol, recreational and prescription drugs and hallucinogens
- intense and prolonged psychological negativity and problems
- bad accidents and traumas

- over-activation and production of the inner fires, called the *kundalini* in Sanskrit, through wrong meditation and breathing techniques. This will be discussed in more detail in Section E on Spiritual Powers.
- psychic attacks
- black magic

These people need immediate help because they are not just energetically open, like a house with damaged insect screens on the windows, exposed to negative entities and forces anytime. They are also confused and outside standard reality. Their ability to hear and feel different worlds is unstoppable. They cannot stop their psychic faculty because the energy sieves and veils are permanently damaged or open, and this is why most drug addicts have hallucinations or paranoia. Visual hallucinations and paranoia are connected to involuntary inner seeing or clairvoyance, and auditory hallucination and paranoia are connected to involuntary inner hearing and clairaudience. Now you can see why they hear and see other realities and inner worlds that other people don't hear or see. Some advanced healing methods to address these issues exist, but practitioners are rare.

Third type: Conscious psychics

These people have been born or stimulated with more advanced development because they are usually fairly safe and correct in their readings, but not so with their interpretation. They know that they have some of the psychic faculties, but they are not knowledgeable about universal concepts and applications yet. The problem they usually encounter is the quality of reading and how to apply the information for service at a higher level. Since there are many levels and ramifications of psychic faculties and their useful applications, this section of the book can divulge untapped potential or their next step. Even though conscious psychics can open and close their inner sight and hearing, they still need proper training from a qualified teacher or group. Before they can find a good mentor, some of them also need to learn humility and need to do more spiritual service to be entitled to more advanced teachings.

Fourth type: Trained psychics

This group has the greatest power and extent of service, if they are guided properly. Some of them have spiritual teachers and groups, hence they are able to receive proper training and have practical applications and opportunities to practice. They can also compare notes on what they experience with other classmates or co-workers. Their spiritual coaches are their best resource of information and validation. If their teachers are practical, they are also trained not only to hold their talents and powers within their group, but are also encouraged to help other groups. These trained psychics can sense and envision some of the global work and Divine strategies ahead of time, thus they are useful for world service.

Flash Forward: Future Teamwork in Humanity

Can you imagine the power of humanity if you combine the accurate foretelling faculties of the trained mystics and psychic seers with the concretizing and scientific powers of brilliants scientists, executives and leaders? We can perceive ahead and solve negative obstacles before they create chaos and destruction. We can also form great teams for even more advanced breakthroughs and discoveries. The real teamwork will not occur in a group of like-minded people or through members with similar temperaments. The future powerful teams and groups will be a combination of "inner seers and external doers". Currently, the mystics, healers, artists, religion and educators seem to enjoy or socialize with each other while businesspeople like to play golf, socialize or work with like-minded people. Scientists are not easy to communicate with, preferring to talk with each other using their own specialized vocabularies. Politicians have their own style of diplomacy and leadership, and most of the time it is not easy to talk to them about long-term goals. But this situation will fade out as the incoming modern generation will have something in common – psychic powers. Then they will form a civilization in which the integration between the objective and subjective workers will create powerful teams. This is the future humanity.

In the past, many spiritual teachers discouraged the stimulation of psychic powers for the safety of those who are just curious and unprepared to unleash the power and also for the protection of victims of the erroneous applications of this so-called sixth sense. Psychic powers are regarded as a paranormal, metaphysical or supernatural faculty of advanced Souls. This is partially correct, but not totally true because based on my personal experiences with many psychics, there are some who are good in mediumship, but many are mediocre and mentally undeveloped. In fact, their minds are still young and the emotional body is quite stimulated but disturbed. Therefore, the purpose of this lengthy introduction is to make sure that aspiring seekers and practitioners do not make their newly acquired or experimental feats a big deal, but proceed humbly to learn more and apply it to help others and society for a better world. With this attitude, they can be trained as good healers, counselors and group members or cultured to become a balanced leader of a service organization.

Fourth requirement: The condition and development of the energy centers

One of the most important factors in the ability to unfold the psychic faculties is the conditioning and development of the energy centers. As we have been studying in the previous sections, the energy centers have not only physiological functions but also psychological and psychic functions, and they act as intelligent data processors. That is the reason behind their ability to sense and absorb, process, store or transmit the information, data or experiences. Let us enumerate the different energy centers used directly for psychic power unfoldment and their psychic and spiritual functions:

Chakras for Psychic Powers	
1. Crown: top of the head	Intuition, higher Divine impressions and universal perception of an idea
2. Forehead – in front of the forehead	Inner sensitivity to Divine impressions and higher clairvoyance
3. Ajna – mid-brow	Clairvoyance, higher psychism and telepathy (especially the mental ajna chakra)
4. Throat – in front of the center of the neck	Clairaudience and its inner concrete interpretation; chakra connected to telepathic work (especially the mental throat chakra)
5. Front Heart – in front of the center of the chest	Emotional intelligence and conscience (especially the emotional body heart chakra)
6. Front Solar Plexus – in the center of the stomach area below the sternum	Lower clairvoyance and sensitivity to lower vibrations and emotions (especially the emotional solar plexus chakra)

7. Navel – in front of the navel	Instincts and spontaneity of physical reactions
8. Sex – in front of the sexual organs	Sexual instincts
9. Base-of-spine – base of the spine or coccyx	Instinct in terms of financial, material existence and physical survival
10. Alta Major – base of the skull at the top of the spine	Receptor and synthesizer of information from the head chakras
11. Back of head – above alta major, directly opposite the ajna	Receptor of information from the ajna chakras
12. Hands – center of the palms	Sensitive to energy impressions, clairsentience and psychometry
13. Right eye – in front of the right eye	Receptor for abstract intelligence (especially at the mental body)
14. Left eye – in front of the left eye	Receptor for concrete intelligence (especially at the mental body)
15. Left and right ears – in front of the left and right ears	Receptors for inner sound and information

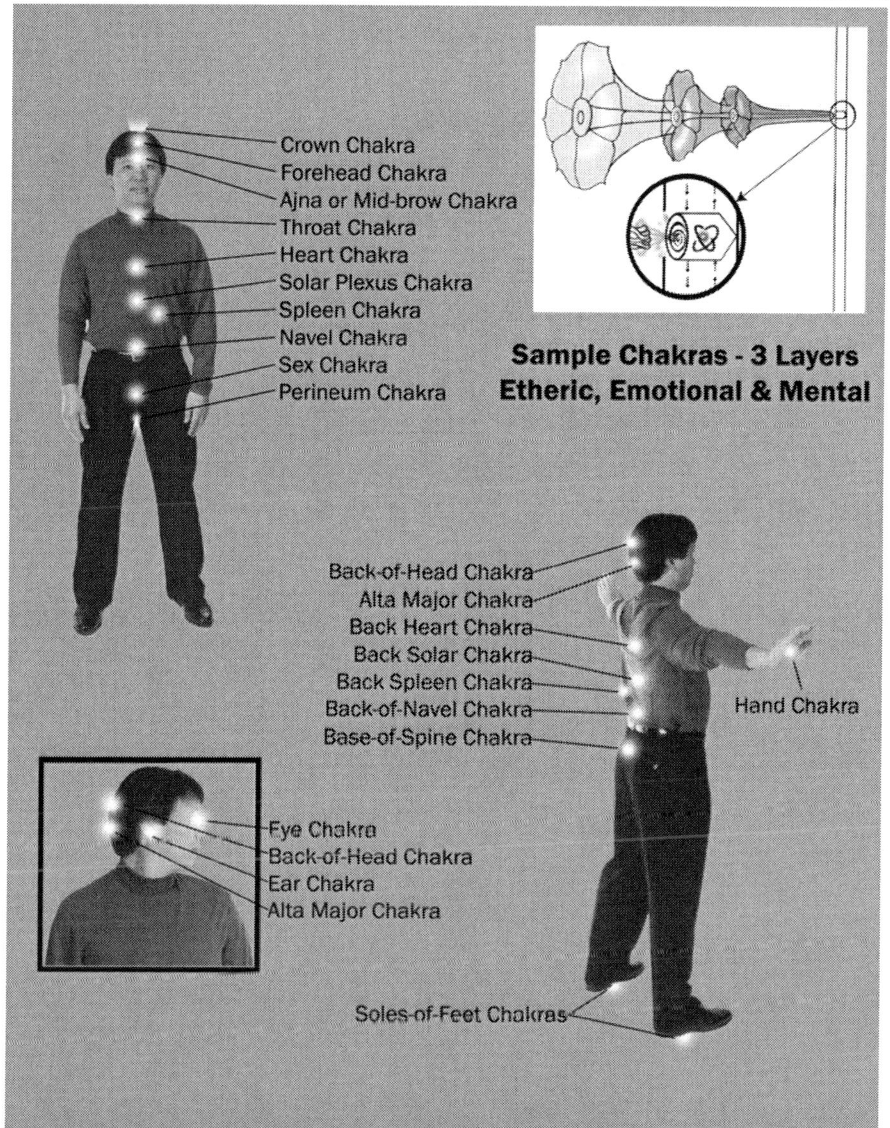

The Locations of the Different Chakras of the Human Energy System, which have Correspondences in the Etheric, Emotional and Mental Levels

Figure 24

As I have shown in previous sections, the chakras have three tiers of frequency corresponding to the three energy bodies: etheric, emotional and mental.

The three levels of energy centers are all interconnected and interrelated – the mental interpenetrates the emotional and etheric counterparts – and they all reciprocally affect each other. It is how clean, energized, activated and integrated they are that maximizes their functions. In most people they are not at all integrated, thus they are either more physically instinctive, emotionally reactive or mentally active. Once a person's chakras are totally integrated as a whole system, then the person's true inner power develops to a very high degree. This is one of the requirements for the integration of the personality. Therefore, there are different levels of psychic unfoldment, from the levels of etheric vision to emotional clairvoyance and eventually mental clairvoyance. They all depend on whether the etheric, emotional or mental body – or a combination of the three – is the dominant perceiving frequency.

Psychic powers

First powers: Clairsentience and psychometry

Clairsentience and psychometry are faculties of sensing the different energy bodies and energy centers and their conditions using either the hands or the auras of the different bodies. This also applies to sensing energies of the environment.

If you are energetically sensitive, you can feel people around you, right? And sometimes you can feel if they are in a positive or negative mood. This is common to many people, especially women and mothers, although some businesspeople and more developed leaders use this sensitivity to gauge their clients and new acquaintances. Some people call it a "hunch" or an "impression" or "goose bumps" or a "gut reaction". Some also use the expression "butterflies in the stomach" if they feel bad about a situation or person. In martial arts, especially internal types like *tai chi*, *chi-gong*, *aikido* or internal kung fu, which employ energy, vitality and sensitivity are important elements in advanced training to sense, with great speed and instinct, the motive and direction of your opponent. This is the key to being able to fight many adversaries at the same time using auric sensitivity called clairsentience. So for the sake of classifying it, one can sense the vitality or emotional and mental auras. There are two types of clairsentience:

1. No-touch clairsentience using energetic sensitivity
2. Touch clairsentience, which is called psychometry

Psychometry is a psychic power of tuning in to the object, its origin or source and qualities. Unlike the no-touch clairsentience faculty, psychometry needs physical contact to link to the subject. This can also be used by a clairvoyant to see psychically what a person is thinking or visualizing. For example, one of my early spiritual mentors, MMCN, being a very advanced clairvoyant himself, uses this faculty. If you ask a question about a person that he does not know, he asks you to touch his hands and think of that person, then he can energetically tune in and evaluate the person's aura or energy centers and discuss the person's problem. So the process is like a connecting cable to transfer the data from the source to the reader or monitor.

In my early spiritual training, I joined groups in the Philippines who practiced psychism. We practiced psychometry by psychically reading the qualities of people with their wallet or handbag. We practiced touching the object, tuning in to the energy and describing the person who owned it. It takes time to master, but some of my classmates who were born psychometrists had very good accuracy.

When I was young, I played basketball and ran track and field, and sometimes I had ankle sprains. I would go to an old man healer for a quick fix: I mean 15 minutes from a painful swollen ankle to no pain and back to the basketball court, especially during basketball league season in the final games. This old man would touch my hand and in two minutes, he would describe the nature of the blockage and damage in the ankle and would tell you that it was no big deal. He was tuning in to my sprain just by touching my hands.

Now some Westerners have even adapted some of these methods. My students in Texas, who are also friends of mine, went to Blissfield, Michigan to visit their folks While there, they ran into an ex-employer who is a veterinarian. Since both of my students are clairvoyants, they were interested in finding out the latest use of energy in veterinary medicine. They were very impressed to discover that the vet uses psychic faculties to diagnose horses. This is what they found: The vet diagnoses the horses through his own pulse. He measures and monitors his pulse away from the horse. Then he goes and stands beside the horse, close but without physical contact. The energy of the horse affects his energy field, which he measures in the change in his own pulse. By that change, the vet evaluates and diagnoses the health and condition of the horse. Isn't that a great display of the alteration of both the person's and animal's aura through energetic osmosis – without physical contact?

Oriental and ayurveyda medicine also use psychometry techniques by counting and the pulse and detecting variations. Psychotronics, a vibrational therapeutic system, also uses evaluation of people's health conditions at a distance by applying psychometry on the person's specimen, like a strand of hair or a nail clipping. I am sure that there are many more applications of these psychic faculties in the future for healing, detection of body conditions and even finding missing people or objects. Some Filipino psychometrists are known for this power. Criminal investigation departments sometimes consult them to trace missing people.

Some sensitive healers can detect a patient's problem because they feel the patient's pain or discomfort in the corresponding part within their own body. This is not the best way to evaluate the patient's disease energetically because the healer can be psychically contaminated, but it is a start. I would like to give you strategies to develop clairsentience as follows:

1. Developing hand psychic sensitivity to feel energetic conditions
2. Methods to use the different auras to feel and evaluate environmental conditions or other people

Procedure 1. Internal martial arts method to sensitize the hands

This is a technique derived from some Martial Arts Masters and groups in China, especially the chi-gong practitioners, to increase your vitality level and the sensitivity of the hands to feel subtle vibrations:

1. Rigorously shake your body, hands and arms like the body shake exercise in the MDP Body Therapy Program. Also flick your fingers several times (1-2 minutes). This is to activate the energy in the body.
2. Stand with your feet apart at shoulder width. Stay in this position until we finish the practice. This stance will improve the ground energy connection to your feet and aura.
3. Place your palms in front of your navel area – both facing each other about 8 inches/20.32 centimeters apart with your fingers pointing to the front. Relax in this position for a few seconds. Breathe deeply 5 times.
4. Rhythmically pull your hands gently outward and back to the original position for 21 sets for about 45 seconds. While

doing it, sense and feel the energy of the palms. Finish with your hands in the original position.

5. Move both hands in parallel upward to the chest level, then slide them down forming an arc until the palms move horizontally outward to the side at hip level (refer to the diagrams on page). Do 21 sets for 45 seconds.
6. Repeat step 4.
7. Maintain the hand position of step 3. Rotate both hands while maintaining them parallel, forming 1-foot/.3-meter circles – in any direction – 21 times. Then reverse the direction for another set of 21 rotations.
8. Again, place your hands in the step 3 position. Relax and breathe slowly.
9. Be aware and feel the energy of the palms while breathing slowly for 1 minute, then very slowly move the palms apart about 3 feet/.9 meters. Then keep your hands in this position for a few seconds, and finally move them closer again until they are about 8 inches/20.32 centimeters apart. Do this slowly and gently while continuing to feel and be aware of the palms for 2 minutes. Feel and sense how your palms react. The key is to relax, breathe slowly and gently and do not fabricate or anticipate results.

Internal Martial Arts Hand Sensitizing Technique
Figure 25

Body Shakes
(1 set - 1 minute)

Starting Position

Shake your whole body to loosen up, and build up the rhythm.

Rhythmic outward hand movements (21 times)

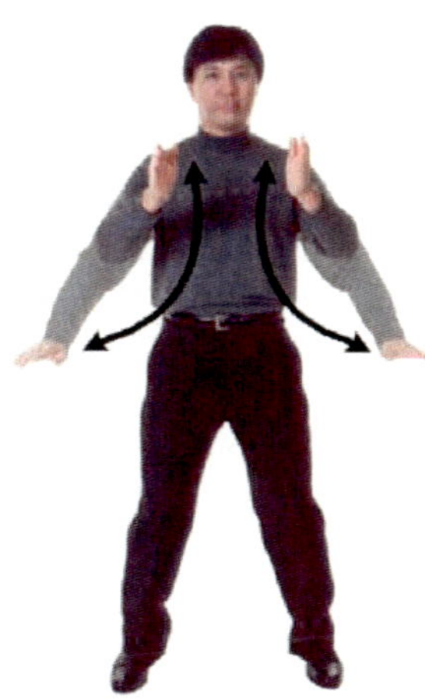

Upward and downward arc (21 times)

Rhythmic outward hand movements (21 times)

Internal Martial Arts Hand Sensitizing Technique
Figure 25

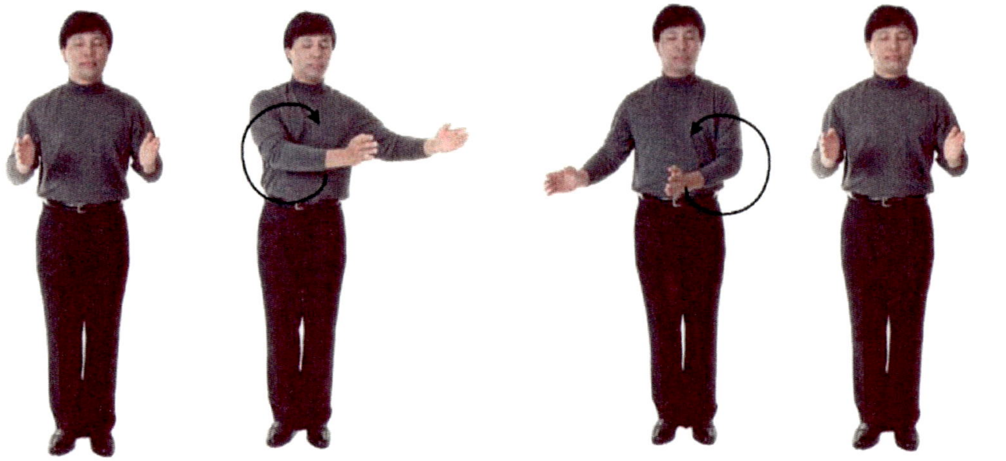

Circular Hand Rotations (21 times Clockwise)

Circular Hand Rotations (21 times Counter-clockwise)

Foreward & Backward Hand Rolls (21 times Foreward Hand Rolls and 21 times Backward Hand Rolls)

Slow Outward & Inward Hand Movement to Feel the Energy (3 sets in 2 minutes)

Did you feel anything? Check if you had any of the following experiences:

- Warmth
- Tingling sensation
- A subtle connection between the palms
- Pulsation or a feeling of opposing magnets between your hands
- Density or heaviness between hands
- An elastic feeling
- Light between or around your hands
- Feeling energized

Record your first experience, and compare it as you do this technique in the future. This initial introduction to the clairsentience faculty of the hands will give you an experience of the vibration or energy that makes up your etheric aura. When this is experienced and recognized, you have transcended the idea that you are only the physical body. Of course, some of you are already familiar with this concept.

Benefits of the hand-sensitizing technique

You can gain several things if you do this technique periodically for 5-7 minutes:

- This technique makes you aware of the concept of subtle energy until it is engrained in your consciousness.
- Increased vitality level and circulation of internal energy
- Activation of the psychic sensitivity of your hands

Procedure 2: Applications of clairsentience by using the hands to evaluate etheric energy

1. Experiment on yourself or another person. Find a painful part of your body that is accessible to your hands, or if you

don't have any pain, ask somebody else around you, like your partner or friend, to be a subject. Ask if they have physical pain or discomfort in any part of the body.

2. Knowing that your hands have an aura, let the aura of your hands and fingers touch the aura of that painful part – without physical contact – with your hands about 3-4 inches/7.62-10.16 centimeters away from the body. Slowly scoop the energy from the aura of the part and sense if you feel a subtle heaviness, weight or change in your palms. If your hands and fingers feel itchy or heavy, that painful part probably is not energetically clean. If it is very itchy or has a prickling pain sensation, that part probably has some negative entities or psychic parasites. Practice evaluating light and heavy vibrations. If they feel light, and there are no reactions from your hands, the vibration is most likely clean or normal. Using your left hand, scoop a sample of energy from a non-painful area and compare it with a sample scooped with your right hand from the painful part. Which energy is heavier? Wash your hands with cold water after the experiment to avoid unnecessary contamination.

Procedure 3: Comparatively sampling the vitality aura for light (higher) or heavy (lower) vibrations and energy

1. With your right hand, gently scoop a sample of your energy at the solar plexus (stomach area) and keep the energy sample in the palm of your right hand.

2. Gently scoop a sample of your energy at the heart center in front of your chest with your left hand and keep the energy sample in the palm of your left hand.

3. Compare the samples side by side to determine which sample is lighter or heavier. Normally the energy from the upper centers is lighter, higher and more refined.

4. Practice regularly to evaluate the refinement of your aura and that of a friend or family member. This is the way to learn – by constant experiments, not blind belief in given concepts. As the hands' psychic powers develop, the palms will become more sensitive to detect information and impressions for more advanced applications.

5. Wash your hands with cold water after the experiment to avoid energy contamination.

Note: While evaluating the energy, use your feelings, rather your mind, to sense since clairsentience uses inner sensitivity more.

Procedure 4: Energy level detection practice

1. Gather two sheets of white paper, which are the same size.
2. Sign your name on one of them.
3. Fold each of them the same way, without knowing which is which.
4. Shuffle them until you don't know which one is signed.
5. Put them flat on the floor 2 feet/.6 meters apart.
6. Start by putting both your palms about 2 feet/.6 meters over the top of the paper, then gradually feel the energy of each sheet as you slowly bring your palms closer to them. Using

the sensitivity of your palms, feel and sense which sheet has a higher level of energy. The higher energy level, which is felt as greater pressure against your palm, is your signed paper.

7. You can apply this technique when you have to decide among 2 or 3 solutions or alternatives. Write the choices on a piece of paper without any biased preferences. Mix the papers up so you don't know which is which. Use the method in step 6 to feel and sense which choice feels best to you. It's usually the paper with the highest energy. Since you have not mastered this technique yet, though, do not use it as your only method of decision-making.

8. You can also do this experiment on different denominations of your currency (e.g., $5, $20, $100). Test which has the highest energy. You can also practice evaluating the energy level of the aura of 250 milligrams of a medicine compared to 500 mg of the same medicine. Which aura and energy level is greater? There are many applications of this technique for business decision-making, but it is not safe to divulge all of them for public use.

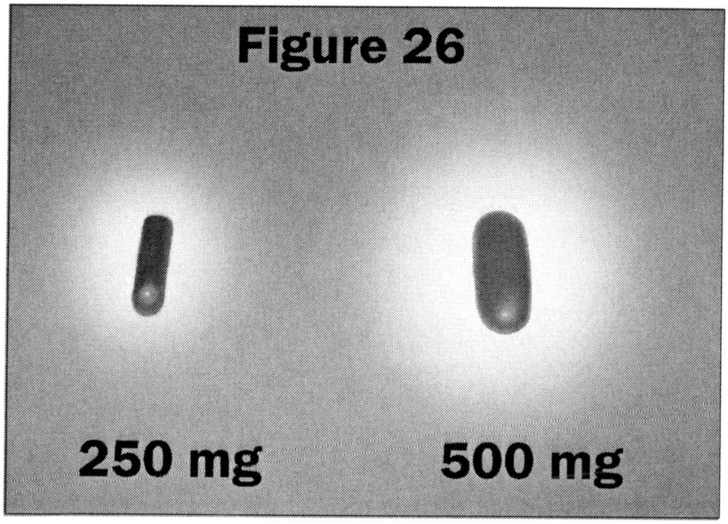

Procedure 5: Practice for feeling and detecting emotional and mental substances and their qualities

1. The steps are similar to **Procedure 3**, sampling of the lighter and heavier vibrations, but this time your intention is to get an energy sample from either your emotional or mental energy body. Use your hands to scoop the energy specimens and compare their refinement or grossness of vibration. Sample yours and another person's. In evaluating other people, you should ask permission if possible or preferably use your family members to start the practice. If you are a healer or counselor, then you can apply it to your patients. Continued practice makes you proficient.

2. If you compare the energy of great personalities like Sir Isaac Newton, Mother Teresa and Thomas Edison or great leaders and world servers, they have more refined and active vibrations compared to most people. The only

difference is that the loving personality types, like Mother Teresa, have a refined vibration in the emotional body and high activity in the heart area, whereas Sir Isaac Newton also has a refined aura in the emotional body and heart area, but the throat and ajna centers are more developed, especially in the mental energy body. This insight is not meant to intrude into people's privacy, but to demonstrate that there is a methodological way to evaluate subtle vibrations, temperament types and behavioral patterns, which are the blueprint and matrix of people's identities. In fact, we use this method as basic training for people to apply clairsentience. While receiving special training with the Pranic Healing organization headed by MCKS, I learned that they are very good at similar methods, including measurement of etheric chakras and scanning the etheric body.

Detecting the sizes of the chakras and energy levels of a person's etheric energy centers and emotional and mental bodies is easier when you do the internal martial arts hand sensitizing technique regularly. Your hands will become like a doctor's stethescope.

Procedure 6: Sensing other people's qualities through auric sensitivity and instincts

This is commonly known as "gut feelings" or instincts. When you deal with new clients or even friends or family members, practice this method:

1. When you first meet people, what is your hunch or feeling about them? Is it positive or negative? Note your first impressions, and compare these findings later when you know the person better. Practice weekly "hunching" as much as possible.

2. Before you enter your house, feel the atmosphere by tuning to your instinctive impression. Compare the moods of the people inside with your reading. You can also do this before you enter your office or meet your friends.

3. Every city also has a different energy field. When you go to a new place, tune in to its behavior as your plane lands or as you drive over the border. Feel the difference in the psychic ambiance.

4. This can also be used to tune in to business deals before you finalize or sign a contract. Sense with your gut feelings, and use your instincts to guide your decision in combination with your normal reasoning faculty.

5. If you are a healer or health care practitioner, note your gut feeling or instinct about a person's condition before examining the patient or client. Assess the individual using your auric sensitivity, and compare your assessment with the complaint or symptoms of the patient or client. Practice, practice, practice.

In all of these practices, very sensitive people tend to empathize too much with the negativity energy of others, the place or the situation. Sometimes you can feel their pain, fears or stress. This is not a good experience. What can you do to disassociate your auras? Just think and will that you are totally disconnected from the person and very far away from each other. You may also think or visualize that you and the person(s) are two separate auras with no connection. Direct your attention to a pleasant situation. Be detached.

Clairsentience is a practical tool to improve your instincts and aptitude for decision-making, especially in situations where time is limited and information is not sufficient to use the process of reasoning. Practice is the key to proficiency.

Psychic powers

Second power: Etheric vision

One of the psychic faculties and powers to be developed is etheric vision or the ability to see the etheric body with your eyes open. The vitality body or aura is sometimes called the etheric body, which is composed of the aggregate energies from air and Earth energy, solar rays and food, up to the higher vitality vibrations from the Earth's etheric aura or from different energy beings. There are generally four degrees of frequencies that make up the etheric or vitality body. Clairvoyants and psychics can see denser or lower frequency vibrations easily, even with their eyes open. Some people can also be trained to see the light of the etheric aura substance with their eyes closed. The good news is that this faculty of etheric vision can be learned and performed with simple methods. Let's demonstrate these techniques with the experiments below

Experiment #1: Seeing the aura of your fingers

1. Exercise your hands and fingers by continuously closing and opening them for 1minute.
2. Do the MDP Synchronized Breathing Technique 2-3 times.

3. Raise your hands slightly above your head and connect the fingers of both hands. Slowly move your hands up to 2 inches/5.08 centimeters apart. Look at the gap between your fingers and breathe slowly for 30 seconds. Don't focus too much. Just relax and close your eyes close slightly. Allow yourself to see the vibration or energy of your fingers.
4. Slowly move your hands further apart, up to 4 inches/10.16 centimeters. Continue looking at the gap between your fingers and breathe slowly for 1 minute. Just relax and allow yourself to see the vitality aura of your fingers.
5. Slowly move your fingers closer together, up to 1 inch/2.54 centimeters apart. After 30 seconds, slowly move them apart again to about a 4-inch/10.16 centimeter distance. Observe the variations of the energy field or aura of the fingers. It is quite normal if you are unable to see much on the first attempt. It's easier to see the aura when you don't use will and mental analysis. Just allow the energy profile to appear rather than visualizing it. You need sensitivity and feeling faculties in order to develop etheric vision.
6. Practice daily for 5 minutes for one week to get good results. Most people improve their etheric vision by practicing repeatedly. Once the brain and the mind become familiar with this faculty and can remind the consciousness about the existence of subtle energy, this feat can develop fully into etheric x-ray vision to see physically through the inside of the hands and body as though they were transparent.

Starting position

Slowly move your hands apart up to 8 inches (about 20 cm) for 30 seconds and slowly bring them back to starting position (2 times)

Seeing Your Fingers' Aura
Figure 27

Experiment #2: Seeing the aura of the head

Have someone stand or sit about 8 feet/2.4 meters away against a white background. It is preferable to use a high vitality or energetic person as the subject.

1. Do 2 sets of the MDP Synchronized Breathing Technique.
2. Close your eyes halfway and allow yourself to feel sleepy or drowsy.
3. Gently stare at the entire person, with more awareness of the head area.
4. As you continue looking at the entire person's profile, breathe slowly and relax – no mental activity or will should be present. Refraining from analysis or expectations of results

will make you more sensitive and receptive to see the energy body or aura.

5. Gently and slowly shift your awareness to the head area, especially above it, but do not pick any specific target of your stare. Diffuse your awareness to avoid concentration. Too much focusing does not allow you to see the aura of the vitality body. You need to relax and be passive to open up etheric vision.
6. If your eyes get tired, close them for a few moments, then gently open them halfway. Then continue practicing seeing the head aura. Be patient and allow the brain and mind to process the newly registered experience. Some people need a few attempts before seeing a glow around the head.
7. Eventually, you will practice looking at the aura of the entire head and body.
8. Practice daily on the head or body aura for 5 minutes to develop etheric vision.

Experiment # 3: Seeing the aura of sunset

At sunset, you can practice looking at the aura around and within the physical sun. The best time is about 15 minutes before sunset.

1. Do 2 sets of the MDP Synchronized Breathing Technique.
2. Look at the sunset, diffusing your sight to the immediate area around the physical sun. Breathe slowly and relax.

3. After 30 seconds, close your eyes. Allow a color to appear inside your head, and deliberately but gently join and merge your attention with the light and inner color. Follow the light and keep merging your awareness with it for 1 minute.

4. Open your eyes and look at the sunset and its immediate surrounding light. Allow yourself to see the alteration of colors. Do this for about 1 minute. If the sun is too strong for you, do not stare directly at it.

5. Repeat step 3 for 1 minute.

6. You can add the following technique to subsequent practice sessions. While looking at the sunset and its surrounding light using the method in step 4, gently focus on your ajna or mid-brow and heart centers and breathe slowly. Be aware that you are utilizing the ajna center as the lens or filter to see through. You will be able to see more if you do this because you will no longer be using your physical eyes to see the light and aura of the sun; instead you will be using your chakras.

7. Practice these techniques once a week for 5-10 minutes and enjoy the splendor of colors and energy revealed through your etheric vision.

Experiment #4: Etheric x-ray vision

Most people will find it easiest to achieve immediate results with experiments 1-3. Method #4 will take more time and practice for most people. You can use this technique to see through the physical body and its organs. Many children have this faculty. According to one of my co-workers in Connecticut, her 7-year-old daughter was watching a video of my MDP Body Therapy exercises and she saw my internal organs and the blood flowing in my body. Since my co-worker is also a trained sensitive seer, she just laughed and assured her daughter that it is a great gift and above-normal faculty. It is very important to guide children with these gifts of sight not to shut them off and to coach them on how to close the inner seeing at will.

The following is a more guided application of the experiments you tried at the beginning of this section on psychic powers.

1. Practice this technique when not disturbed by people or the surroundings because you need a peaceful, quiet ambiance.

2. Do 3 sets of the MDP Synchronized Breathing Technique.

3. Feel and be aware that you have an x-ray or infrared ability to see through objects or people, that everything can be penetrated and internally photographed. Be aware of this concept as you practice this method. It is important to establish this concept in your consciousness to work without mental blockages created by the limitation of needing the tangibility of physical matter.

4. Look at the refrigerator door and penetrate the inside of the refrigerator as if it were transparent. Explore the contents inside. Be passive and patient so that you allow the profiles to appear as flowing imagination rather than

visualization. Do not use the mental concentration of enforced intention or will-power. Register any information or profile in your consciousness.

5. You can also practice this with your house. When you are outside, penetrate the contents and movements of people inside. You might believe that it is just your visualization or imagination that gives you the profile or image, which is partially true for starters. But as you do it with unfamiliar places, like a hotel or a city, you will have more practice and results based on the etheric x-ray vision faculty. You can even try this when opening gifts. Before you open the box or package, use your x-ray etheric vision to see inside first.

6. Also practice looking through your hands, fingers and feet. Some of our trained students have become excellent seers using x-ray vision. They are able to see the movements of the heart and blood when they have high vitality.

The higher techniques of etheric vision are not allowed to be presented in this book, but are taught in workshops or special coaching and inner training. The advanced techniques involve the use of the different centers and are fertilized and stimulated by the sacred fire, the *kundalini*. They also involve activating the energy centers, opening their "shutters and lenses" and remove the webs or the energy sieves to permanently see through. These techniques are supposed to be practiced only by advanced students when their different energy bodies and centers are already substantially purified and transformed.

In the future, when etheric vision, especially x-ray vision, is taught in schools and universities, especially in the medical and healing fields, the doctor and healer will be able to see the internal parts of the physical and the etheric bodies to detect their abnormalities for fantastic corrective and preventative health care.

Most diseases express themselves first in the etheric body and its chakras before they manifest in the physical body as symptoms or diseases. Some advanced healers can perceive diseased energies through the organs' energy centers before the patient can feel the symptoms. Sometimes a person may be experiencing pain or symptoms and they get a medical check-up. The doctors don't find anything, but the pain or symptom persists or periodically reappears. This happens to energetically sensitive people who feel the problem through the negative condition manifesting in their aura and energy centers. It takes a certain period for the physical materialization of the disease. That's why some of the health practitioners think patients are "psychosomatic", saying, "It's just in your mind". On the other hand, some people can produce diseases by believing that they have the physical problem, which mentally programs the disease to appear through the power of their mind.

I mentioned before that I have a vision that the blind can be taught how to see and read using their etheric vision. They can do this by using the energy centers or etheric body to see, bypassing the physical eyes. To medical scientists and most people, this is not possible, but I have witnessed a few people in Indonesia who are able to read words and names and find objects while blindfolded. This was also witnessed by hundreds of people at a spiritual convention in Jakarta, Indonesia in 1997. This internal martial arts group also tests their students by having them navigate around obstacles while blindfolded. They were able to do it successfully as if they were seeing with open eyes – believe it or not! As I have mentioned earlier, many incarnating children are born with these faculties, so let's hope that parents and schools will be ready to teach and guide them without fear in a nurturing and understanding environment. If not, most of them will be home schooled or become aloof to the traditional learning environment.

When a scientist is trained or born with the faculty of etheric x-ray vision, perceived reality will be much deeper than what electronic microscopes can see or what our physical instruments can detect. Higher truths integrating the objective and subjective realities will be discovered or revealed. The missing gaps in truth and phenomena will be resolved and seen properly and correctly with truer perspective. When many people are able to see the now so-called "invisible light" physically, there will be a big revolution and breakthrough in spiritual, educational and scientific consciousness, and this will mark the rapid advancement of a different type of human evolution.

Psychic powers

Third power: Clairvoyance

The faculty of clairvoyance sees beyond the inner realities of the etheric bodies or worlds. The clairvoyant is also able to see the emotional and mental energies and inner worlds, with eyes open or closed. This type of psychic uses the emotional or mental energy centers or energy bodies to see the inner light, energy forms, profiles and existences of the subjective worlds. The chakras, which are the centers of energy in different bodies, have an aperture or opening in them that is usually thick in most people, but which can be opened by trained or born psychics, using it as a shutter lens or windows to peep through. There are three layers of energy sieves or shutters. The inner two are the most important for psychic development. As we have mentioned earlier in this section, the "chakral shutters or lenses" are able to open in some psychics who cannot control their faculties or are not technically trained. But they need more coaching to understand what they perceive and to be able to apply the inner reading to some useful service. We also discussed the damaged energy sieves or screens in substance abusers that allow them to see the energies and inner world permanently without control, understanding or protection. In this state of clairvoyance the person experiences visual or auditory hallucination and paranoia, which that cause them great and possible insanity.

The Different Conditions of the Chakra's Energy Screens/Sieves Figure 28

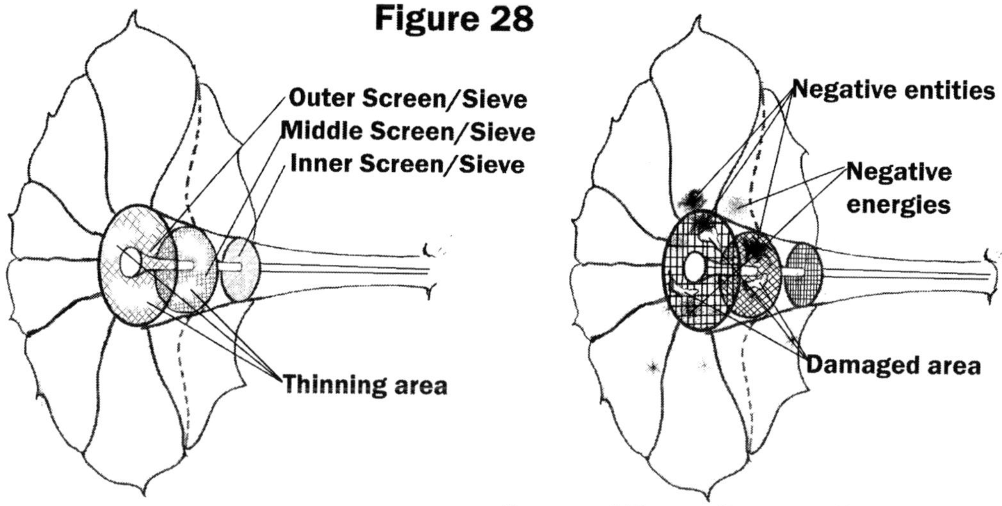

Thinned Energy Screens/Sieves

Damaged Energy Sorccns/Sieves with Negative Energy Contaminations

Therefore, clairvoyance results from the ability to open the chakral energy screens in order to see through them as energy lenses. If the sieves are damaged or only very thin screens remain, they become transparent, which also leads to clairvoyance. The ability to peep through the holes or thinned layers also allows people to see through to the other side of the screen into the non-physical frequencies. I mentioned earlier some methods whereby the protective energy sieves are punctured, damaged, psychically burned or removed. The methods employed here are simple and basic techniques, which are safe for beginners. Special courses are designed to teach people clairvoyance depending on their energetic purity, inner development and motives for using the acquired psychic powers. Gaining more inner powers is a responsibility for both the teacher and the student. Therefore, the additional information and faculties should be used for greater service and should serve as an important ingredient for spiritual development. The following are some basic techniques to practice clairvoyance.

Experiment #1: Candle flame practice

Chakra development and brain stimulation:

1. Light a candle and place it at least 8 feet away from you, preferably in front of a white background. It is best to practice this in the dark.
2. Do 3 sets of the MDP Synchronized Breathing Technique to purify, revitalize and empower your energy bodies and energy centers.
3. Close your eyes. Focus on your crown, forehead and ajna energy centers and breathe slowly and deeply 10 times. Do not use strong concentration or will-power. Be relaxed and maintain gentleness throughout the experiment.
4. Be aware of the alta major at the base of your skull and breathe slowly and deeply 10 times. This step enhances your ability to see the energy better.
5. Be aware of the heart center and breathe slowly and deeply 10 times. This technique makes your feelings more sensitive.

6. Be aware of your brain and imagine it to be inside a bright ball of pinkish-golden light. Breathe slowly and deeply 10 times. This step is required to revitalize the energy of the brain to register the incoming information and experience better.

7. Focus on your navel center and breathe slowly and deeply 15 times. This step allows you to produce and absorb more vitality to sustain a higher energy level.

8. Now you are ready to practice seeing clairvoyantly. First, look at the candlelight for 2 minutes with your eyes open, but diffuse your physical vision to also capture the light radiated by the flame. Do not focus or concentrate while looking. It is better to defocus or half-close your eyes while looking at the candlelight.

9. Close your eyes and look as if you were still looking at the candlelight. Allow the gentle appearance of an energy profile. Do not visualize using the mind. Be patient and wait for the formation of the images or light. If you cannot see anything after 2 minutes of waiting, open your eyes again.

10. Look at the candlelight and breathe slowly and gently while simultaneously focusing slightly on your heart and ajna energy centers. Relax and defocus your eyesight so that you are focusing on the light radiating as a whole rather than on the flame or candle. Continue to practice even if some of you don't have immediate results. Constant repetition and practice give the energy centers and the brain sensitivity to perceive and register the new

experience and concept. Continue the slight focusing on the heart and ajna centers while breathing slowly and gently. Look at the energy images of the candlelight while your eyes are closed. Just relax, be patient and practice. Allow it to happen and be sensitive. Don't be willful. Also avoid mental analysis. Don't evaluate yourself while practicing so that you avoid involving the mind. Clairvoyance does not involve logical or concrete thinking.

11. Repeat step 10, but focus slightly on the heart and forehead centers.

12. Repeat step 11, but focus slightly on the heart and crown centers.

13. Practice 10 – 15 minutes daily for a week and a few times a week thereafter. Some people will have immediate results and will be able to achieve the initial stage of clairvoyance. Some might be able to see the inner light and aura of the candle flame clairvoyantly, which is different from the physical light. Others might be able to see some strange variation of light for a split second. A few will probably see some inner forms of light expansion. And some people won't be able to see the aura of the light or any variation of the physical light easily. Everyone has different faculties developed or stimulated, so whatever your experience, you gain something because your brain, energy centers and etheric, emotional and mental bodies are starting to register experiences, information and concepts never recorded or perceived before.

Experiment #2: Advanced method to see the auras of the sun at sunset

The physical sun has etheric, emotional and mental auras. Therefore, the sun can be the subject of clairvoyant investigation. Using the energy centers as the mechanisms or tools to see the inner light of the sun is quite a different experience for clairvoyant practitioners.

Repeat clairvoyance experiment #1, but use the sun at sunset as the target instead of candlelight. Do it once a week, 10-30 minutes before sunset. Record your experiences in a clairvoyance record notebook or spiritual diary. As you note your experiences, your subtle realizations become more vivid. Writing in your spiritual diary or clairvoyance notebook helps you stimulate the memory aspects of your mind and trains it to recall and register more subjective impressions.

Experiment #3: Seeing the energy bodies and energy centers of human beings

Highly energized people are easier to see clairvoyantly than devitalized and tired people. It can be draining to look at a depleted or sickly subject clairvoyantly. When you are viewing the human energy system clairvoyantly, you can see through different chakras. The solar plexus is the entry point or "telescope" to the lower emotional world. It can see mostly negative and lower types of energies, forces and inner forms. Ordinary mediums or psychics generally use the solar plexus. They need to shift their development and stimulation to the upper lenses and "inner world telescopes" from the ajna, forehead and crown. Using the crown to see through, one can perceive the more refined worlds, forces and higher vibratory light and beings. The ajna is a seeing lens at the mid-range vibratory levels, a higher frequency than the vibrations seen through the solar plexus. The forehead can see the vibration and energies in between the range of the ajna and crown centers. Therefore, the four centers are the most common seeing energy windows or telescopes in clairvoyance. The heart center serves to enhance sensitivity, overcoming will and concrete mental power and bringing more stimulation to the brain of the seer.

The alta major at the back of the head approximately at the base of the skull and the back-of-the-head chakra are your photo films or monitors to register signals and information of different images. We mentioned earlier that the energy centers are like the lens of your video camera. The purity and cleanliness of the energy center determine the clarity and quality of the captured pictures. The purity and quality of the auras of the seer also determine the integrity and quality of the interpretation. Therefore, the MDP Synchronized Breathing Technique is very important as a purifying and revitalizing technique.

Now you can practice looking at a person's energy bodies and chakras:

1. Do 3 sets of MDP Synchronized Breathing Technique.
2. Gently focus on the navel center and breathe slowly and deeply 15 times.

3. Be aware of your heart center simultaneously with your ajna, forehead and crown energy centers and breathe slowly and gently 15 times. Developing clairvoyance to see different energy bodies requires stimulation of the different chakras of each energy body: the mental chakra for mental clairvoyance and awareness of your emotional body chakras for emotional clairvoyance.

4. Gently focus on your alta major center and back-of-the-head chakra, and breathe slowly for 15 times.

5. Imagine a person in front of you or an actual person if available and look at his or her aura with your eyes closed. Wait for images and inner sight to emerge. Just relax and be sensitive. Do not use will-power and mental analysis. If it doesn't register, don't be frustrated. Your "inner video" might be seeing it already, but mental and brain recognition and registry are not stimulated yet. It takes some degree of processing and practice to tune up the brain and consciousness to capture subjective realities that have not been realized and recognized before, especially for very mental and willful people. Use the emotions and feelings to become gentler in your process of unfolding your clairvoyance.

6. Do steps 1 and 2 a maximum of 2 times in every practice so that you don't over-stimulate your energy centers. Practice seeing the auras as in step 5.

7. Practice 20-25 minutes 3 times for 1 week, then a few times weekly thereafter until you achieve better results.

8. You should receive higher clairvoyance training from a qualified teacher through either a seminar or personal coaching.

Experiment #4: Seeing the auras of animals, plants or natural crystals

Most animals have vitality and emotional bodies, and a few have started to form a very small outline of the mental body, especially domesticated animals. Plants also have a vitality aura and are starting to have a small outline of the emotional aura, especially some fragrant flowering species. Minerals like crystals, diamonds and precious stones have small but grosser frequency vitality auras. The auric light and intelligence are the substances that make some, like healing crystals, capable of absorbing and implementing simple energy instructions. The light consciousness at the etheric levels of frequency in crystals is sufficient to program instructions and information for them to follow. You can use the clairsentient faculty to feel and sense the energy fields and levels of the crystal, plant or animal. This method was discussed in the previous section under clairsentience and auric sensitivity.

Practice this method #4 of clairvoyant development with an animal, plant or crystal as a subject:

1. Do 2 sets of the MDP Synchronized Breathing Technique.
2. Gently focus on your navel center and breathe slowly 15 times.

3. Gently focus on your heart and solar plexus center and breathe slowly 10 times, with more emphasis on the heart center.

4. Gently focus on the ajna center and breathe slowly 10 times.

5. Practice seeing the aura of any of the above subjects. Be patient and maintain your passive gentle mood. Relax in order to enhance energy sensitivity (5-10 minutes).

Experiment #5: Clairvoyant practice with distant subjects

One of the most important applications of clairvoyance is for investigation and evaluation of a subject or event at a remote or distant location. The subject can be evaluated from the past, present or future. This requires more training and skill. The idea is to look at the subject as if he or she is in front of you, not afar, with specification of the time, date or period required for the evaluation. If the location is known, then it is easier for the clairvoyant to interpret other parameters. As mentioned earlier, the energy of the environment affects the reader and the subject being read. There is a possibility that the seer can be affected by the ocean of thoughts and emotions floating in his or her immediate environment or in that of the subject. This is where many psychic readings at a distance result in a discrepancy of results. Here is a basic process and practice:

1. If the subject being read is a person, follow the guidelines of clairvoyance experiment #3, steps 1-5. Look

clairvoyantly using the heart and head centers during the breathing practice.

2. If the subjects are animals, plants and crystals, use the procedure of clairvoyance experiment #4, steps 1-5.

When clairvoyance is substantially mastered by healers and medical practitioners, the higher and future conditions of the health of people can be predicted and monitored more accurately and rapidly. Even genetic conditions and health tendencies of human beings, animals and plants will be monitored and evaluated through clairvoyant faculties. The culturing of civilization will not be left to a long arduous suffering, trial and error method of refining human beings, but will involve a rapid upgrading and transformation of humanity to produce enlightened cultures. Clairvoyance is one of the major keys to this future success.

Psychic powers

Fourth power: Clairaudience

This faculty allows the clairaudient person to hear inner sounds, emotional signals and mental thoughts partially or completely. Some of them can also listen to emotional and mental messages and communications from other people or invisible beings. My students report numerous cases where they are able to hear people's conversations even at a remote distance. They call these people to verify the subject matter and to their surprise, their evaluation of the conversations is very accurate. This is the normal level of clairaudience. In more highly developed clairaudience, the person with this gift can hear (not through the physical ear) inner music and sounds from the invisible world like sacred sounds and music from the angelic realm or higher sacred worlds. Some of the composers that I have talked with sometimes experience their musical composition being guided by inner musical dictations. I believe some geniuses had some faculties connected to clairaudience or similar faculties like telepathy (the faculty of reading messages or non-verbal inner communications mentally) when they were children. We will discuss telepathy later in this section.

The problem is, both good and negative messages and emotional and mental influence can also be communicated to people with clairaudience faculties. We have discussed examples of auditory hallucination sufferers like drug addicts who hear inner voices from the inner realms. To these people the voice and communication are as real as if you were hearing with your physical ears.

The mechanism behind clairaudience is the energy centers of the ears and the throat centers, especially if the energy sieves or screens are partially or greatly damaged or ruptured. Therefore these shutter mechanisms of internal hearing are opened without a noise insulator. The inner world noises, vibrations and communications are able to be received through the throat and ear energy centers. This is not just in the etheric body but especially in the emotional and mental levels. In fact, one of the functions of the throat energy center is to perceive inner information and process it through analysis, especially the mental throat chakra. The emotional throat center is more for the perception and reception of the emotional energies, whereas the mental throat chakra is for mental and thought substances. Therefore clairaudience is very much allied with or similar to telepathy.

People born with this faculty or those who are clairaudient because of damage to the energy sieves of the ears and throat centers can be contaminated by the involutionary forces or negative entities when they are in negative states like anger, hatred, fears, greed, selfishness and violence. Negative energy in human beings attracts these negative influences. These types of people, especially those who have damaged ear and throat centers due to excessive drug and alcohol use, can be used by negative beings to implement wrong and destructive plans that are harmful to themselves or others. Most of them do not even know that they are being used.

This can explain why kids who take recreational drugs can be used by an inner voice, or what Christians call temptation, to commit crimes or even kill their own family members. Some women who are depressed with post-partum psychosis can also fall victim to negative inner voices and dictations forcing them to commit suicide or unexpected crimes. Many criminals and extremely violent and negative personalities suffer through these influences without knowing it, nor their families or counselors know it.

Research and evaluation into these phenomena for greater understanding and correction can help avert negative events. Healing and psychotherapy techniques already addressing these conditions do exist. These conditions are more familiar to most psychotherapist or doctors as paranoia or hallucinations, but I am very positive that solutions do not just involve chemical levels of treatment, but should also focus on the etheric, emotional and mental energy bodies and their energy centers. Treatment must include their proper repair and energy rehabilitation.

There are a few techniques to develop the ability to tune in or use the clairaudience faculty, not necessarily to hear inner messages from the inner world, but to improve the ability to receive and interpret the incoming information. If you already have clairaudience, these techniques will show you how to use it for good service.

Training requires being able to close the faculty of clairaudience when you want to shut off your inner hearing. Listening to too much inner noise, talking and voices continuously can be tiring, stressful and frustrating, especially when you hear the negative thoughts people project to you. Most people with the clairaudience faculty wish they did not have this power because they are frequently disturbed by inner messages, both good and bad. The purpose of the following method is not just to develop your inner skills, but to purify the energies and energy centers responsible for you to interpret received inner messages or information so the proper meaning and interpretation will be achieved and not corrupted by the negative vibrations which might be present in your energy centers.

Method to enhance clairaudience or inner hearing

1. Do 3 sets of the MDP Synchronized Breathing Technique.
2. Gently focus on your throat and heart energy centers and breathe slowly and deeply 15 times. This will purify, revitalize and activate the inner reception of the energy centers. This technique also improves your interpretation and perception.
3. Gently focus on the ear energy centers and the alta major center at the base of the skull and breathe slowly and deeply 10 times. This is to improve the "spiritual microphone" and the "spiritual monitor" for clarity of reception of the messages, information or inner signals.
4. Verbalize the sacred words OM, Aum or Amen 7 times. These universal words are used by most religions as sacred words. The recitation of OM, Aum or Amen produces higher frequencies and vibrations that elevate the energy of consciousness. After reciting the words, be relaxed and quiet.
5. Practice your inner hearing by listening in your consciousness to the inner sound of OM, Aum or Amen

that you just recited. Do not mentally produce the sound. Let the sound OM, Aum or Amen appear or be replayed in your inner consciousness. Relax and wait patiently for the inner sound for 2 minutes. Avoid using mental analysis or will-power.

6. Do steps 4 and 5 again.

7. Practice this experiment and method 10 minutes daily for a week and periodically thereafter.

This technique will not only train you to become a better receiver and interpreter of ideas, but it also gives you a higher grade of premonition or the ability to recognize future events. Clairaudience is a cousin faculty of telepathy. Sometimes they can be interchanged. Many so-called clairvoyants are more of clairaudients because they don't see the visuals immediately, but they hear answers or the information before they recognize the inner data through visuals. Many people also channel unusual information and claim to be visionaries. Most of them are the clairaudient type who are hearing messages or dictations rather than receiving visual information through clairvoyance. There is always the danger of receiving wrong information from a negative source if the clairaudient is in an energetically dirty environment or if his or her energy bodies are not purified. Many are just attracting or listening to psychic garbage in the environment.

The main uses of the clairaudient faculty are for tuning in to the advanced forms of music, sounds, sacred words, *mantrams* or higher ideas to help guide the evolution of humanity or the Earth. I believe that many great musical composers, artists, advanced visionaries and prophets have used this inner hearing faculty to bring down masterpieces or ideals. Another incoming use, which is already used by Saints and Sages, is to apply this inner faculty to communicate with the Angelic beings, or what we call in Sanskrit, *devas,* or invisible helpers. The Divine strategies of the Great Ones always work with the help of these inner world builders of forms. They construct the matrix of the future world. When you utter words to create a goal, the *devas* construct it into an energy form, which eventually materializes as results or events.

> *Deva* or angelic evolution runs parallel to human evolution. When the conscious integration of the two groups becomes common, the gap between objective reality and subjective realm will disappear, and this will eventually result in continuity of consciousness for humanity.

Direct communication with the Angelic Kingdom will allow many important universal principles to help human beings bring new truth and information for the incoming stages of human development come to life. This process will cut out many primitive research tools employing a trial and error method, which are too slow and expensive and are subject to many errors and wrong conclusions. Therefore many advanced clairaudients who will become authors will be tapping into the higher inner world of ideas and universal "news" to bring down all this super-advanced information to complete the missing levels and gaps of Earths evolution. Most of the advanced great composers and authors will become intelligent and selfless transcribers of pure truth. When the lower aspects of the personality and obsolete qualities of these potential channels are purified, the purer type of clairaudients will rise to great usefulness with unthinkable data. The future is just around the corner…

Psychic powers

Fifth power: Telepathy

Telepathy can be classified as a lower or higher psychic faculty. In the higher form, the telepath receives mental thoughtforms or ideas. This is what we are interested in learning in this book. The lower form involves reading lower emotional forms and feelings psychically. Clairsentience and clairvoyance can substitute for this lower faculty so we will focus on the mental type of telepathy in order to avoid duplicating techniques and training.

Telepathy is the ability to sense, interpret and mentally process ideas, thoughts and mental forms that come to you. It also includes the ability to create and project thoughtforms and clear messages to be sent to others or groups for good purposes. It uses the following chakras to receive and process the information, especially the mental energy centers:

1. Throat center
2. Ajna center
3. Forehead center
4. Crown center
5. Alta major center
6. Back-of-the-head chakras
7. Sometimes the ear chakras
8. The mental counterparts of the brain

Since the mind has two aspects and seven vibratory frequencies, we can classify mental telepathy still further as:

- **Concrete mental telepathy**: Concrete mental telepathy uses the four lower frequencies of the mind that compose the concrete mind. Telepathic interconnection, transmission and interpretation occur at this level. Therefore, this faculty intepprets received concrete ideas, normal thoughtforms and telepathic communications. Receiving a thought to call someone or tuning in to another person's intentions or motives are examples.

- **Abstract mental telepathy**: Once a person has already developed and put the abstract mental faculty, which uses the highest three grades of mental frequency, into active use, abstract mental telepathy is possible. This telepathic faculty and application is a more advanced power to receive and interpret or interconnect and transmit more advanced and refined abstract ideas and information not common to ordinary thinkers and telepaths. This is what makes abstract mental telepaths more rare and advanced than people with concrete mental telepathy. The abstract mental telepaths can process very advanced and futuristic ideas that have not yet descended into the realm of human thinking, concrete analysis and common sense. Therefore, these people who are able to tap into this abstract world of ideas in the higher planes of the inner world are the true advanced thinkers. They are the pioneers and visionaries. They become the world leaders or are leaders of pioneering projects in their own field of expertise.

  Abstract mental telepathy requires a more refined mental energy body and high speed, developed chakras that produce mental vibrations at sub-atomic or atomic frequencies in order to expand to the Soul consciousness level and to the abstract mental plane of the Earth. This Soul plane is where we can tap into the world of advanced ideas. Therefore it requires a higher degree of inner development and awareness. This is only possible if the sacred fire called the *kundalini* has stimulated the abstract mental section of your energy system. Thus, your *kundalini* level of development directly relates to your faculty advancement. The subject of the *kundalini* is discussed in Section E on the Spiritual Powers.

There is another side to mental telepathy. The ability to discern if the information via a thoughtform or message emanates from a good or bad source. Why? Because there is no need to bother with the message if it is just garbage or an obsolete or low type of data that will confuse or pollute your mind. Therefore, you need to know if the telepathic data or thoughtform coming into your aura and attention are:

- a pollutant from the mental environment of your local area
- a message sent to you by your family, relatives, friends or normal personality acquaintances
- a message from your business associates or co-workers
- information from your Soul to your personality
- an instruction, guidance or information sent by your inner teachers or spiritual guides
- a planetary message given to the Earth or humanity for a bigger purpose

To be able to perceive the incoming higher thoughtforms or messages with abstract mental telepathy, you need higher auric sensitivity than with other methods of clairsentience or clairvoyance. You also need to purify the different energy bodies and their energy centers to make them as clear a film as possible to trap the incoming data, idea or thoughtform. Also, if your vibration is high and clean with no lower personality pollutants and environmental negative frequencies, the interpretation will be more accurate with no personal or environmental bias. The more refined and higher the frequency of your mental substance and energy centers, the higher the ideas you can perceive. Therefore, it is necessary to receive and read in an environmentally cleaner and purer environment to practice abstract mental telepathy. For normal daily applications of concrete mental telepathy, you can do it almost anywhere when you are required to use it for practical reasons.

Abstract telepathy also requires substantial availability of spiritual frequency and energy from the Soul to the mental energy body in order to illuminate it to perceive and understand the abstract idea. Telepathy is not just the ability to receive or even perceive the incoming message, but also involves the proper interpretation of the information, including where the idea originated. As with TV, radio or cellular phone signal transmissions, the more advanced and powerful the receiving satellites or dish, the better the reception and fidelity of the data received. It is the same in telepathy: The mental chakras should be highly active, refined and developed. This mental faculty is enhanced by the Soul's spiritual energy. The alta major center has important roles to help imprint, perceive, process and integrate higher data. The higher energy centers, through their built-in intelligence, absorb and scan the data at different levels.

- The crown center is used for higher-frequency data with a formless interpretation.
- The forehead center receives moderately high-frequency data with a mixture of Soul frequency reception and personality frequency information.
- The ajna center is for average-frequency data with a more abstract idea or thought.
- The throat center is for more concrete frequencies, working with normal personal thoughtforms and data accompanied by auditory perception and interpretation.
- The ear centers are auditory receptors.
- The back-of-head chakra is more of a visual receptor.
- Lower psychics use the solar plexus to receive lower types of messages and information.

Another higher work to be assumed by the telepath is to be able to create, send and respond to messages or thoughtforms, if it concerns actions in group work. These techniques will not be taught in this book, but for people who are doing this for higher group work, you can form the intentions or information with visualization and send them as a wish to the concerned person or group. This can also be perceived as telepathic communication.

Do you want to practice developing your telepathic powers? Let's do it. The thoughts and mental wishes of people can be transferred to others through telepathic rapport and idea transmission. Therefore be careful if you are deciding amongst people with conflicting ideas. You will be negatively affected and will get confused. Through mental intention, withdraw yourself and your aura from others if you want to have your own clear perception and decision without their biased influences. It's even better to physically move out of the room if possible, disconnect yourself from other people and think in your own mental space. It's better still if you know how to shield your auras.

Practice 1: Conditioning your telepathic equipment and developing your concrete mental telepathy

Procedure

1. Do 2 sets of the MDP Synchronized Breathing Technique.
2. Gently focus on your heart, throat and ear energy centers, especially at the mental body level, and breathe slowly and deeply 15 times. This is to activate the receiving and perceiving centers of your body.
3. Gently focus on your ajna, forehead, crown and alta major energy centers, especially at the mental body level, and breathe slowly and deeply 15 times.
4. Be aware of the whole concrete mental body while breathing slowly and deeply for 2 minutes. This is to activate, sensitize and revitalize the mental body to perceive incoming information and messages.

5. Do this 3 times weekly to activate the telepathic ability of the concrete mental body and to enhance the receiving and perceiving energy centers.

6. Whenever you are waiting for answers or decisions from any member of your family, friend or co-workers, try to use your mental sensitivity or telepathic perception to tune in to their wishes. Practice tuning in by being passive in a relaxed, waiting mood until certain interpretations come through your mind without creating them.

7. Also practice tuning in and guessing who your callers are telepathically before you answer the phone. Many people use this instinctively, and it has become a normal experience for them. They may know who is calling on the phone before picking up the line (not aided by the phone's caller ID feature!). It's good practice for simple telepathy.

8. If you are relaxed and peaceful and suddenly you feel some emotional or mental discomfort for no apparent reason, just be quiet and passive. Tune in first to who is sending you information or the negative feeling or thought. Is it yours or somebody else's? If you think it is from someone, the first person who comes to your mind is the potential sender. Mentally test it by rejecting your guess about the person. If his or her image comes back again and even stronger, then your guess might be correct. The message is from that person. Tune in to the contents of the message. What do you think the person wants you to know? Is it urgent? Is it a need for telephone call, or does the person miss you and wishes that you were there?

Tuning in is not that difficult, especially if you already have a good telepathic connection with that person or group. Practice whenever possible and call the person by phone to verify and evaluate the accuracy of your telepathy. Always be objective and non-fanatical about what you are interpreting. Do not always assume that your readings are the rules of truth. Properly investigate to prove your accuracy. Don't take it too personally. Be humble even if you are already an expert.

Practice #2: Conditioning your telepathic equipment and developing abstract mental telepathy

1. Do 3 sets of the MDP Synchronized Breathing Technique to purify, revitalize and upgrade the frequency of your energy bodies and energy centers.
2. Concentrate on the navel and breathe slowly and deeply 15 times.
3. Gently focus on the heart, throat and ear energy centers, especially on the mental levels of the chakras. Breathe slowly and deeply 15 times, then be aware of the whole concrete mental body and breathe slowly for 2 minutes. This is to energize and stimulate it for greater reception and sensitivity.
4. Gently focus on the ajna, forehead, crown and alta major centers, especially at the mental body level, and breathe slowly and deeply 15 times. For more advanced practitioners, be aware of your abstract mental body and breathe slowly. This is to activate the abstract mind to

connect to the higher world of ideas. Imagine the abstract mind as a higher rate of vibration enclosing the concrete mental body or just know yourself as the Soul while breathing.

5. Mentally sound Om, Aum, or Amen 7 times, vibrating it on the concrete and abstract mental bodies. This increases the rate of vibration of the mental body and integrates both levels.

6. Remain calm, relaxed and silent for 3-5 minutes.

7. Invoke for higher ideas or messages connected to your spiritual life's purpose or group spiritual work. If you have a spiritual teacher or an affinity to great Saints or Sages, you can invoke them to guide you with an enlightened idea or strategy for problem-solving and important decision-making. Your Soul is also a ready source of inspiration or noble ideas if you do not have teachers or guides. Just relax and wait for an impression. Do not preempt insight by visualizing. Do not use mental analysis or evaluation until impressions comes through. Let ideas or impressions precipitate without visualizations or mental preoccupations.

8. Whenever an idea comes, let it flow. Do not interfere by editing or analyzing it. Writing it down later or even drawing a symbol are the best ways to register incoming information. Abstract ideas can be physicalized not only by forms of concepts, but also by symbols or quotations or maybe poems.

9. If nothing much happens, be patient. Sometimes when you sleep and your mind is not very demanding and analytical, the idea comes to the mental body and as soon as you wake up, there is a new input. Write it down immediately without editing the data. Sometimes the brain has no terminology or concrete interpretation for a new unfamiliar principle. The Soul's inspired idea or higher spiritual message can stay in the mental body for some time until there is a sufficient intelligent correlation to interpret it.

10. Make sure the data really came from a positive source. To validate it you can ask a more advanced person like a spiritual mentor, coach or another clairvoyant to validate your perception. Another test is to analyze the idea or concept to see if it does not violate universal values or principles of goodness, altruism, benevolence and fairness or ideas that do not contradict the Golden Rule and violate the Law of Cause and Effect. Evaluate new ideas and principles you tapped into for coherence with existing valid truths and principles. It should not create separativeness.

This method and self-training can be used even by the experienced psychics to improve their mechanism of perception and accuracy of their interpretation.

The real higher purpose of telepathic training is to accelerate the distribution of higher ideas from Divine Macrocosmic sources to qualified recipients for future global strategies so that they are not contaminated and corrupted by the limitation of languages or concrete expressions in order to:

- Be able to see the contents of the idea without linear, limited and inhibited expression through words. Some more advanced concepts are unable to physicalize because they have correlation to existing phenomenon or realities. To express them in words is tantamount to desecrating the idea. Only through telepathy can they be expressed completely.
- To enhance education and deeper inner learning experiences that do not belong to normal mundane reality. This is done as a process wherein the higher ideas are being planted in humanity to replace obsolete, mass opinions and to upgrade common sense at global levels. New universal ideas and strategies need to be disseminated globally without the slower process and edited version of commercial communication and media. Advanced and futuristic information is generally rejected if improperly spread through normal channels like the media. TV viewers instinctively and emotionally reject and negate new revolutionary concepts without proper and deeper analysis of their rightness and usefulness. The information through telepathy does not get colored by the personality of the ordinary spokesperson, media or religious advocacy.

There are great beings, usually called the Great Ones in the esoteric schools and also known as the Communion of Saints or the Society of Illumined Minds who, in their high places and development formulate and send these universal ideas and Divine strategies that are components of the organized plan on Earth. The vision and direction of the future world and destiny of humanity and all kingdoms of nature are guided by these highly evolved groups. Most of them are not in physical bodies, but some are walking in our cities. The tapped ideas, principles, global projects and evolutionary schemes – including political, religious, philosophical, business, scientific, educational, arts and music and positive endeavors – are all guided by these Great Ones. They inspire, stimulate and empower all the great leaders, pioneers and icons in these fields of human development through telepathy via their advanced students. The pioneers, advanced thinkers and mystics have, in one way or the other, connected to their intentions and Divine plans, bringing down progressive, creative and global ideas that eventually become physical realities. Therefore, those with more advanced abstract mental telepathic development have been their instruments and channels to implement global projects and evolutionary ideas.

Several writers and spiritual students of these great enlightened beings have been used through telepathy, clairaudience, clairvoyance or a combination of these faculties to transcribe or channel advanced spiritual teachings. If you observe most religions, their holy scriptures were not written by their founders, but I am positive that the writers were inspired and utilized psychic faculties and inner powers to physicalize advanced teachings and the scriptures. The first Western spiritual disciple to bring spiritual books in our modern time was Madame Helena Petrovna Blavatsky who, in 1888, authored *The Secret Doctrine*, which was a major source for most subsequent metaphysical books and esoteric information. The writings of Madame Blavatsky also formed the major basis of the teachings of the Theosophical Society. Until now, many spiritual seekers are benefited by the deeper philosophy and principles, said to have been given through telepathic dictation by three Holy Beings whose initials are M, KH and DK.

The second body of Esoteric Teachings from the Ageless Wisdom, the common name of the inner teachings from the Group of Illumined Sages and Masters, came through Alice A. Bailey, the author who used a combination of telepathy and clairaudience to transcribe 23 volumes of books, first published in 1923, through 30 years of telepathic dictation by a Tibetan Sage, Holy Master DK. This advanced Sage was also responsible for dictating part of

Madame Blavatsky's book. The last volume of the Alice Bailey series was completed before she died in 1953.

The books through Alice Bailey form a large body of advanced esoteric teachings, which have accurately guided and also interpreted many world events. The books are so abstract that even very intelligent scholars can hardly understand or penetrate their meanings and principles. Many of the teachings in the books on *Inner Powers* are an elaboration or an extension of these earlier esoteric works from the Ageless Wisdom. These books on *Inner Powers* explain and express through experiments and practical applications many of these universal teachings, but they are more focused on the science of human inner development and culture of civilization.

I believe that even the great ideas captured by scientists like Albert Einstein, Madame Curie, Thomas Edison and the post-World War unification of countries through the United Nations and other global projects were inspired and empowered by the Divine Plan formulated by the Great Ones. It is common esoteric knowledge that the Christ, called by many names in different religions, and two other great Beings, Lord Manu and Lord Mahachohan, lead this inner group of Advanced Sages.

Today, many great leaders, icons, pioneers and authors are unconsciously tapping into the world of ideas that will bring in the new strategies for the incoming great age of humanity, using telepathic communication and other psychic faculties.

Psychic powers

Sixth power: Precognition faculty

This faculty is sometimes associated with premonition or the ability to sense or predict future events. This is quite common in many countries, especially in cultures that are more emotional or mystical in nature. Many people use this faculty through their dreams. They experience seeing or dreaming the events before they happen. For instance, some of my mystical clients dream or suddenly see a flash of vision that a relative of theirs or friend will die soon – and they get scared when their premonition comes true. They start having a phobia about their dreams or vision. The problem for them is how to determine whether what they precognized as incidents are authentic or just a product of their dream state or imagination.

As I was writing this section of the book on the psychic powers, something happened to amaze me even though these faculties of clairvoyance and precognition are real and my group has used these inner powers extensively for healing and inner spiritual work. The Twin Towers incident in New York City on September 11, 2001 caught my attention because it fulfilled a premonition some of my students had several years earlier. One of our students from Caracas, Venezuela is a clairvoyant who came to New York for a healing seminar in July 1997.

Our group went to Weehauken, New Jersey across the Hudson River to view the Manhattan skyline at night with all its light and splendor. While we were looking at the whole stretch of the city, this psychic student suddenly became hysterical, crying so uncontrollably I had to ask her what happened. It took a few minutes before she could talk. She saw the buildings downtown burning and collapsing, approximately where the Twin Towers were. She saw in a flash of a vision the explosion and many deaths.

On September 4, 2001, a few days before the Twin Tower incident, one of my students from Bogota, Colombia visited New York City for a meeting, and while she was in the lower Manhattan area, close to the World Trade Center, she almost physically felt the suffocation and partially experienced what happened in the burning buildings. She was so affected by the experience that she immediately called one of our healers in Connecticut to help her recover from the shocking precognitive experience. She was with us in Texas for a seminar during the attack of the Twin Towers, and she almost collapsed in shock when she heard the news. The incident of the burning buildings was very similar to her precognized experience a few days earlier.

Many of my co-workers and students have also reported the precognitive faculties demonstrated by some of their children. I am sure many more will encounter these experiences of having a vision, dream or precognition of incoming events or situations. Nostradamus was said to have this faculty. Even in the Bible and many other religions' ancient histories, one or more personalities and prophets always recognized big future events and possibilities. The premonition or precognition faculty is also associated with being a visionary.

The psychic power of precognition can be an admixture or combination of clairvoyance, clairsentience and telepathic faculties that recognize the future films of events ahead of time. Before things physicalize, they are already reality in the inner world. The writing of part of Volumes II and III of *Inner Powers* used these psychic aptitudes to bring the potential matrix of events in human development from the future to the present in order to be studied as guidelines of possibilities.

How do we apply this faculty? When you sense or see a flash of vision or incoming events that is not visualized by your mind, but is a continuous parade of ideas or incidents, you may do the following:

Precognitive psychic faculty development technique

Procedure

1. Write down the vision or premonition and analyze it to see if there is a connection to what you programmed as a thoughtform or a desire or the wish of someone around you. You might be reading your or someone else's mental imagery or thoughtforms. Also, the thoughtform can be actively sent to you by someone.

2. If it is a dream or vision of a negative incident either concerning you or others, cancel the image or vision of the incoming event by willing it to disintegrate or disappear. Do this several times with strong will-power. There are many advanced techniques to do thus, but it requires special training.

3. If you think that there is a lesson for you or the concerned person to be learned, study and go through the lesson as soon as possible. If the lesson is learned earlier, the possible problem or negative event might be averted. If you want to be sure, donate a reasonable amount to a good cause corresponding to the difficulty or possible lesson you need to overcome. Dedicate the good karma of your donation for the problem or possible undesirable conditions to be prevented or minimized. Try experimenting with this so that you have your own experience and conclusions.

4. The precognition faculty is automatically developed when you have developed clairvoyance, telepathy and clairaudience. In the future, the predictions of psychic faculties will be combined with scientific instruments to

> predict natural calamities or negative incidents. This faculty of prediction will plan the future better for individuals, groups, businesses and nations.

Another practical esoteric method to apply this psychic skill is to evaluate a person's chakras to predict future tendencies. When a psychic sees clairvoyantly or using clairsentience that your base-of-the-spine chakra is shrinking and very depleted, you might not feel the condition immediately, but he or she can predict that in a few days, if it continues, you will be physically depressed or tired. If this goes on for months or years, he or she can also predit that your finances will go down. By knowing the patterns of the chakral and auric conditions, you can also graph the future possibilities in your health and life. This is a more sequential conclusion rather than mystical. These methods will be used in the future to forecast people's health and psychological conditions by reading the patterns of the human energy system.

It seems that Latin America, the Philippines and Southeast Asia have good psychics. This is probably because of their mystical and more emotional nature. Some of my students in Central and South America have been good at telling me ahead of time about things that actually happened later. The only factor hard to pinpoint is the exact date, so normally there is a range of periods – like months – during which events can be expected. This is the variation in predictions read by different channels – the time element. The energy patterns and probable events or incidents are already designed in the energy field of the Earth or in a person if it regards an individual, but the time frame is not absolutely fixed.

In these modern times, precognition can be useful to predict certain world events and Earth physical changes before they happen. I also believe that certain happenings are Divinely veiled so that people will go through sometimes challenging situations in order to maximize their lessons in life.

This explains why even the best psychics cannot see much of their own future or how to become very rich instantly.

Psychic powers

Seventh power: Channeling, automatic writing, drawing or movement

This psychic faculty has been observed clairvoyantly to employ two methods of channeling. The first is channeling given by another overshadowing being, and the second is the fecundation and inspiration by the Soul. The first, popular in the mystical world, is overshadowing by another being, incarnated or discarnate, upon the physical and vitality body of the person who writes or draws things automatically. The person is not conscious or does not deliberately move the hands, fingers and pen. In fact, most of them are not even aware of the drawings or symbols formed or the article written unless they study them later. If the subject matter is very abstract and deals with advanced topics that promote evolution and support the progress of life, there must be an advanced Guide or Being overshadowing the automatic writing faculty. If the person automatically writing channels negative topics that are menacing to human development, then his or her overshadowing guide is an involutionary force or being.

When I taught a course in Quito, Ecuador in 1997, one of our participants brought her 15-year-old son who is considered to be a gifted artist and poet. At an early age he would automatically draw many artistic symbols and paint colorful works of art. He also created many poems by writing them without thinking. He just produced drawings automatically by moving the pen. I talked to the boy and evaluated his psychic power. I found out that he usually feels like a force of energy is using his hands to move the pen or combines with inner impulses to write different poems.

As I scanned the drawings, artwork and poems, I discovered that there were two forces using him. He did many drawings of unicorns and nice paintings, but the poems were created by negative guide because they focused on a combination of violence, sex, fears and anger. I saw the potential danger in him that his faculty would not be used positively. His mother requested healing for him, so I removed and disintegrated negative entities planted in his energy centers that attracted involutionary forces to use him. Three months after the healing session, his mother told me that the subject of his poems shifted to topics on spiritual virtues and that the drawings changed from darker red and orange colors into more green and violet hues, including several angelic forms.

These examples are provided because many youth today are very creative and artistic, and some of them probably have automatic writing faculties. One of my students in San Juan, Puerto Rico, a 22-year-old, is very psychic. Sometimes he utilizes clairvoyance, but when he has a pen in his hands, he automatically draws without thinking. I ask what it is and sometimes he doesn't know. It can range from a very abstract shape and symbols to concrete shapes. When he finishes, I usually see meaning in them like an energy center of the body, an entity of the place or an invisible being connected to fire elements. At first, he did not understand or have technical knowledge about the subjects of his drawings. My job was to explain them to him using the concepts and wisdom from Esoteric Sciences. Now he understands most of what he channels.

The second type of automatic writing is guided directly from the Soul of the person or by the abstract mental intelligence, which bypasses the brain or concrete mental body. Why does this happen? Usually, the concrete mind using the brain only sequences or processes information about what it already knows and has learned before through the five senses and influenced by general common sense. Since a new, channeled idea cannot be comprehended logically, the information is channeled directly by the higher mind, bypassing the normal thinking process, which then controls the physical and etheric body to write or draw automatically. This proves that the brain is not the main controller of everything that we do. It is an instrument of the mind and the emotions to implement many things. But as humanity advances, the brain and even the emotions and logical mind, can be bypassed. The Soul and invisible beings can utilize the etheric body and the nervous system together directly to move the body without a person thinking, even if the movements or dance have not been experienced or learned before.

Another 26-year-old student of mine in San Juan, Puerto Rico is a great belly dancer. She belly dances professionally, but when she is highly energized and inspired, she dances without an archetypal format. Her automatic movements are "channeled" and guided by her Soul and another special invisible force. Sometimes she is amazed at how the very difficult but graceful movements come about – without planning or knowing them. That is why she does not normally duplicate the same dance steps and forms. I observe that she is dancing with a bright golden light directing her movements while she just follows the internal force and flow. Therefore, this type of psychic faculty is accompanied by clairsentience or sensitivity to subtle energies.

I have also experienced channeling directly. When I started teaching martial arts, I discovered by teaching it to others, that there are limitations of each style however good the lineage is – probably because of the founder's preferences. Having studied at least 6 martial arts systems from different origins and styles, I got bored with what I knew and what was available in books. Since joining competitions had enthused me to improve myself further, I had to innovate many fighting moves and art forms to be better than the opponents. I finally reached a saturation point of using the best movements and techniques. At that time, I instinctively invoked the greatest masters of martial arts. For a few days nothing happened; there was no new, channeled power. I was taking moves and creating them, so it was still an innovation or revision of the old systems. My will and mind were too strong, and I was still the one controlling my actions. Since I got frustrated with no results, I quit expecting anything and surrendered to the fact that I wasn't getting anywhere.

One day, after exercising for two strenuous hours, my body got a little bit tired and became passive. Before I left the gym, I decided to practice some of my karate forms for a few more minutes. I started and suddenly my body wanted to move differently. I felt the need to flow and surrender, and for about 15 minutes, new creative techniques and movements came as a complete advanced style. I used the new fighting form for a few years and it helped me win in both fighting and form competitions. (Form competitions are pre-arranged sequences to demonstrate a fighting technique.) After that incident and channeling experience, I learned that to channel or bring new creative ideas, visions and plans, we need to set aside our mindful control over what we know and what we like and just surrender and be patient to wait for the incoming guidance or guiding force. Sometimes we have to forget what we know.

Many advanced teachings, sacred dances, poetry, music and arts will come to humanity through advanced channels and automatic writers, dancers and artists. The channel needs to be able to discriminate and know the source of the information, guidance and the force that delivers the impulse to him or her. The quality, intentions, contents and universal coherence of the material resulting from the automatic channeling determines the authenticity of the source, whether from positive or negative forces or beings.

Not all people can have automatic writing, drawing or dancing faculties, but if ever you want to start developing the faculty or enhance your already developed faculty, here are a few strategies to improve it:

Practice on Channeling

1. Physical exercises, especially the MDP Body Therapy and Inner Power Development Program, are recommended because they purify, revitalize and balance your human energy system and the energy centers. Follow the steps in Part II, Section A, page 54 or use the video for this technique.

2. Do three sets of the MDP Synchronized Breathing Technique. This will purify, activate and upgrade the energy bodies and bring more Divine light into the system.

3. Invoke for Divine guidance, protection and the right project, ideas or movements when you perform the automatic channeling. You have to be in a clear psychic environment with no negative or contaminated people around you. Do not force it – but maintain yourself in a passive and surrendering mode. If it is ready to come, it comes. If not at that time, be patient to try in another time.

4. Do channeling when you are only in a positive state and high energy level. The quality of information and the ease of doing it are affected by the quality and level of energy that you have. It is preferable to do it after a good meditation.

5. Burn sandalwood incense and use this to purify the immediate environment. It is more powerful than a candle.

6. If you are in a sitting position, it is best to seat facing East. This allows more Divine light to flow with this direction. If you are lying down, your head should be pointing East.

7. Chant sacred words like OM, Aum or Amen or *mantrams* that increase your auric vibration and that of the environment. This technique attracts higher guides or overshadowing beings.

8. Make sure that the channeling is done for a creative or positive goal and purpose. Good intentions attract positive guides.

9. If you are able to channel and perform automatic writing or drawing – use discrimination to evaluate the information afterwards. In the case of automatic dancing or movement, make sure the channeled activity improves health and enhances the quality of your life; it should not be destructive or negative.

As you advance in your inner journey, your Soul and overshadowing guides can periodically inspire you with lofty ideas or creative principles and projects. It is normal that you will enjoy this faculty, but don't forget the power of the mind. You still need the mental powers to concretize new ideas. Therefore, combine all possible faculties and inner powers that you have learned earlier to maximize your performance and improve the quality of your life.

Psychic powers

Eighth power: Telekinesis or psychokinesis

The psychic power of telekinesis is rare and hard to prove because the really powerful people who do it are very secretive about their faculties. There are generally two types of telekinesis or psychokinesis:

1. **Energy telekinesis**: the ability to move invisible forms of energy by mind intent.

2. **Physical telekinesis**: the ability to move physical objects without physically touching them. This still uses invisible force.

I have tried both energy and physical telekinetic techniques, and the energetic type is easier to prove even though moving physical objects is a fantastic demonstration. The concept in this type of psychic development is to first know and understand the energy mechanisms of the aura and the invisible forces. The first concept to know is that energy can be converted into force if you give it directions by applying will-power and strong intentions. Since everything is made up of energy fields invisible to the naked eye, which can be felt or psychically seen, then with your vitality energy or using your hand's energy field, you can push, move or direct things using the available or created force. Because the invisible energy field of everything is interconnected with all-permeating light substance of different frequencies, the mind intent creates the force that moves first the energy of the object, then eventually the physical object itself. For the sake of simple training and practice, I will give you two experiments to prove this principle. More advanced techniques are not permitted to be explained in this book. Here are simple but experiential methods to develop telekinesis:

Experiment #1: Energy telekinesis

We will experience creating an energy ball of light in our hands and moving it to another hand or location using our mind intent or mental power. There are two things that we can prove here: the ability to create energy balls and the power to move them with mind intent and mental force.

1. Stand in your feet at shoulder width. Relax.

2. You may do the sensitizing your hands technique for 5 minutes to make your hands sensitive and to increase your vitality (see page 211). Energy telekinesis is easier if you have high energy.

3. Put your hands in front of you parallel to each other about 12 inches apart. Relax and breathe slowly and deeply. Continue the inhalation and exhalation with the intention to form an invisible ball of white light between your hands. Do this process for 3 minutes, then compress the energy ball to a more compact invisible ball, the size of a baseball.

4. Place the energy ball manually over your left palm and put your right palm beside your left.

5. Feel the weight of the energy ball and compare the feeling of the right palm that does not have the energy ball. Which feels heavier or has a downward pressure?

6. Knowing that the energy ball is on the left, instruct the energy ball to go to the right palm. The best method is to think of the destination - the right palm - and intend for the energy ball to go there. Wherever your mind is focused, that's where the energy goes. This is the true expression of "energy follows thought or mind intent." How does your right palm feel compared to the left? Is the ball transferred? If yes, then this is energy telekinesis.

7. If you want to move the energy ball back to the left palm, use the same procedure. Think of the left palm as a destination and with your mind intent, move the energy ball. You do not need to be willful. Conserve your power by relaxing. Just mind

intent is good enough. Did you feel the difference on your left or right palms? Whichever hand has the ball feels heavier.

Starting position | **Forming an energy ball** | **Place the energy ball in the left palm** | **Instruct the energy ball to go to the right palm**

Energy Telekinesis
Figure 29

The key to energy telekinesis is first to be aware that energy exists with a substance made of light and that it can be moved by mind power since everything, including the energy ball and your mental body, is connected. The mental force through mental intent will direct or force the energy ball to move. The most important guideline is to focus on the target, not the source.

One of the applications for energy telekinesis is to send positive healing energies to people, your home, groups and organizations even at a distance. How do you send the energy of love, harmony and peace to your family, friends or co-workers? Here is a simple application and strategy to help harmonize yourself with them.

distance. How do you send the energy of love, harmony and peace to your family, friends or co-workers? Here is a simple application and strategy to help harmonize yourself with them.

Telekinesis technique to bring love, harmony and peace

1. Do 3 sets of the MDP Synchronized Breathing Technique.
2. Gently focus on your heart and crown energy centers and breathe slowly and deeply for about 1 minute. Feel love, joy and peace as you breathe.
3. Continue step 2, but form the intention of creating a few energy balls containing the substances of love, harmony and peace: The color in your imagination for the balls is pinkish-golden. The balls should be about 6 inches in diameter. Do this for about 3 minutes with pleasant feelings.
4. Think of your family as a group or friends or co-workers as a group and instruct the energy ball to bring love, harmony and peace to them. You can also do this to people and organizations that need transformation and healing. I recommend 3 times weekly for families or groups.
5. You can also send these energy balls to your home or office to bring more love, harmony and peace.

When you create these balls, make sure you program that only good energies and intentions be sent. This technique can improve your relationships with others and in the group. When love, harmony and peace substance are present, it is a fertilizer for the auras of people to produce and experience these positive qualities. It also fulfills the law – what you give is what you receive back. Positive intentions and energies beget positive energies and events. I encourage you to experiment and validate this principle for yourself. You may also send these energy balls when you hear about people in need on the news. The technique can be done at a distance. Ponder on the result of the previous experiment – the energy ball follows your intention and thought. You can direct and send them anywhere in the world, but only with positive intentions.

Experiment #2: Physical telekinesis by pushing a person through his/her aura

This technique has been used in internal martial arts to prove the power of their systems. It is also used by advanced martial arts teachers to defend themselves at a distance when absolutely required. The Pranic Healing organization of MCKS has used this as an experiment to validate the existence of inner powers. One of my other early spiritual mentors, SHJ, also uses this if he does not want to be disturbed by people when he meditates.

Procedure

1. Do 3 sets of the MDP Synchronized Breathing Technique to purify and revitalize your aura. The more refined you are, the stronger your power.

2. Do at least 20 squats to increase your will and vitality power.

3. Let a subject stand 3 feet away in front of you and tell the person to close his or her eyes, be neutral and not to resist.

4. Touch the vitality aura of the person with your two palms and gently feel the person's energy. You need to be relaxed

to do this. Just hold your outstretched hands about 2 feet in front of the person – until your feel the person's aura like a pressure against your palm.

5. With your hands' energy force, push the person's aura backwards and have the expectation that he or she has already been pushed and is moving backward. Do this a few times. The last time, sustain your hand position fully stretched towards the person. Combine both intention and the energy push. Be calm when you do this to conserve your energy.

6. Practice steps 4 and 5 for 3-5 minutes. Observe the effect on the subject. Does he or she get affected by the energy push? When you pull your hands away, does the person get affected, leaning forward or backwards? Ask for feedback from the subject regarding what they feel (pressure or movement, for example).

7. This training includes clairsentience faculty development. Practice daily for 5-10 minutes.

This inner power and telekinetic faculty should be used for internal energy training and not to hurt anyone. It has been included for the purpose of increasing energy awareness and demonstration of psychic powers for greater future usefulness. Several yogis and teachers in Southeast Asia have told me about physical telekinesis – moving objects using the mental and vitality powers – but physical telekinesis won't be discussed here because there's no safe use for it at the moment. Anyway, you have seen this a lot in movies. The concept remains – whether moving objects, subtle energy or people without physical contact. It is the power of energy and intention that makes it work.

During full moon, when the energy of the Earth is highest for each month, people absorb and circulate more energy and inner power in their bodies. The environment also has more circulating energy at this time. This is the reason why many of my students tend to switch electrical equipment on and off around them without physical contact – just by passing by them. Many of them, when totally recharged with vitality, also experience setting off car alarms from 10 feet away. I have experienced this many times myself, but since the principles are understood, it becomes common sense. I'm sure it's familiar to you when somebody is planning to surprise you from behind. You can feel that a presence creates pressure from behind, and you look back to discover your friend creeping towards you. Children are especially sensitive to these experiences.

Psychic powers

Ninth power: Emotional and mental body projection

This subject has been quite discussed in some books and articles as astral projection. Some writers also talk about projectology or dream control techniques. In this book, we will discuss not only its significance, but also the general mechanism behind it. The human energy system is composed of different bodies:

1. Physical
2. Vitality or etheric body
3. Emotional or astral body
4. Mental body: concrete mind
5. Soul bodies starting from the abstract mental frequency
6. Divine Body related to the Spirit

The Soul bodies and Divine Body will be discussed in more detail in Section E on Spiritual Powers. We have discussed the nature and functions of the bodies of the personality – the physical, etheric, emotional and mental constitution of the person – in previous sections. When a person sleeps, they normally go out of body. This statement or expression is commonly accepted by most people.

What does it mean to go out of body? It means to say that when a person is totally asleep, the physical and etheric bodies stay in bed, while the emotional and mental bodies exit and go somewhere in the inner world to do something else.

Therefore, the energy bodies of emotions and thoughts are still linked with their light cords to the physical and etheric bodies but they are not enclosing and fecundating them. The body is still alive and functioning governed by the etheric body, its energy centers and its own automatic physical intelligence and program. This explains why the emotional and mental bodies of the person are still working – feelings, thinking and expressing relationships with other beings in different frequency settings – within the emotional and mental energy fields of the planet Earth. The dream state operates under these conditions. This gives meaning to the experience of people meeting others in their dreams who are not familiar to them in physical life. People also talk about "déjà vu". This is an expression in which you are meeting someone for the first time, but it's as if you already know the person. This also occurs with familiar places that you've never visited before.

You might question: If your eyes are closed and your brain is asleep in your bed, why you can see, move, fly and go to places known and unknown in your sleep and dreams? That is the function of the emotional and mental energy bodies – for you to be able to travel or explore while you are asleep when the physical body does not limit them. This also explains why people can perceive future events or what we studied earlier as the precognitive psychic faculty. In their sleep, some people can see the future through the matrix of energy that is already real in the inner realm. As we mentioned earlier, anything in life first has an energy pattern before it physicalizes as events. Calamities such as earthquakes or volcanic eruptions that happened in 1999-2000 were perceived normally beforehand by some clairvoyants normally in the inner energetic patterns of the Earth's energy field, before they happened physically.

The purpose of the emotional and mental energy bodies leaving the physical and vitality body is:

- To allow rapid recovery of the physical structure without the interference of the mind and emotions. The physical body, through the etheric body, undergoes more rapid repairs and rehabilitation without the two other bodies. In many more advanced human beings who are used for bigger energetic work, their etheric body is also being rejuvenated by invisible healing beings while they are asleep.

- To release the emotional and mental energy bodies away from the limitation of the physical form in order to study and experience the emotional and mental realities of the Earth – without the limitation of the brain and physical location. In very advanced human beings, their inner travel extends beyond the planetary realm to the world of the inner cosmos to further their higher studies. The problem is that negative people with very toxic and low frequency energies who also go out of body while sleeping are traveling to the inner world dominated by the involutionary or darker forces and beings. They bring back unpleasant and involutionary ideas and plans that are destructive and harmful. This is why we need training and education for people (including children) to go out of body only to the right positive inner places.

A proper procedure is available on how to go to the inner world – in your sleep or dream state – with a higher vibration in order to travel only to the higher frequency and heavenly states of the inner world. We will discuss this procedure in a moment.

What happens when people physically die is about the same as people who go to sleep. Their emotional and mental bodies exit and leave the physical and etheric body. The only difference is the light cord connecting the Soul to the physical and etheric bodies is permanently severed, so there is no more life or spiritual electricity to make the physical and etheric bodies function. The Soul now at this point uses the emotional and mental bodies as its instrument in the inner world, but still within the emotional and mental energy field of the Earth. What we call hell is the lowest frequency of the astral or emotional body of the Earth. It is the grossest vibration of the Earth containing all the negative energetic trash or low frequency beings and emotional bodies that are deposited and trapped there for a long period for purificatory purposes.

Purgatory is a low frequency and vibrating energy field of the astral Earth containing and accommodating toxic and negative emotional bodies and entities – low types of emotional vibrations, but not as negative and gross as those of the hell state. Therefore, the energetically gross, dirty and toxic humans who physically die cannot go to the higher frequency states of the inner world because there is a screening of frequency or a threshold for every inner world state or colony. The more negative the beings and vibrations, the deeper and lower frequencies they go to inside the Earth and emotional energy planes.

In the higher frequency of the emotional inner world lies what we call the "first heaven" or a little bit brighter emotional Earth inner world. The concrete mental planes of the Earth start the second heaven of the world – or just a brighter, more pleasant and intelligent frequency where more mental and good people who die dwell. It is where the mental body floats and resides after releasing and dying from the physical, etheric and emotional bodies. So there are different levels of inner world experience after death or when a person leaves the physical body and sleeps. Many people who meditate also leave the physical body, either to go to the inner world or to ascend their consciousness into the realm of the Soul. The different planes of the Earth will be explained later in Section E on Spiritual Powers.

I would like to hint that this emotional and mental projection faculty is a psychic activity of most human beings almost every time they sleep. I say almost because when a person is comatose, their emotional and mental bodies are out already and only the life thread supplying the physical and etheric energy body is linked and maintained by the Soul or sometimes due to artificial medical means like a respirator. Generally, the mental body is not connected to the brain anymore. There lies the big distinction between the mind and the brain. They are closely associated with each other, but not the same. The brain can be alive, but the mind gets disconnected from it, resulting in the comatose state. In the case of trance, the emotional and mental bodies leave the physical body in a totally deep sleep-like state.

In order to connect the phenomenon of getting out of body to your normal experience, let's review what many people have experienced when they go to sleep:

- Feeling of falling into a deep abyss. Most of the time this is caused by the emotional and mental bodies leaving the physical body.

- Kicking or jolting with the legs – a feeling that you are ejecting out

- A feeling of subtle vibration of the physical and etheric bodies moments before the emotional and mental body leave the physical body

- Seeing yourself lying down in bed as if you were on the ceiling. This shows that your emotional and mental bodies are hovering above the physical and etheric body but not totally projected yet in other planes. This situation is also experienced by many people who have accidents, are sick or even by those who are meditating. Sick people think that they are already dead because they see themselves lying down in bed while they float above their physical body, so a lot of them claim this as near-death experience. As long as the life thread connecting the Soul to the body in the etheric heart energy center is still linked, the person is not dead yet. In the case of severe accidents where a person suddenly becomes unconscious, they also experience this out of body state of hovering above the physical body.

Going Out of Body to the Inner World
Figure 30

There are several recommendations for individuals to do before they sleep if they want to maximize their inner experiences while sleeping or during what they call dream state:

Maximizing inner experiences during sleep technique

1. Purify the physical and etheric bodies with physical exercises.
2. Do the MDP Synchronized Breathing Technique 3 times to purify and revitalize the human energy system and activate the heart and especially the crown energy centers. The best exit when people sleep or physically die is through the crown center. It is the portal to the higher and more refined worlds.
3. When you go to sleep, silently and continuously chant the sacred *mantra* OM, Aum or Amen while being aware of the crown energy centers. This will further activate them and allow you to exit easily through this inner door of consciousness.
4. For best results sleep with your head pointing East. This direction also is the best for people who are about to die because the crown energy center is more active and the influence of the Divine energy is stronger than in other directions.
5. If you want to join different activities in the inner world, as you sleep you can also invoke for spiritual guides to accompany you to special events or places that are useful

for your life and inner development – like inner study or research of more advanced philosophies or technology. I am positive that many of our great leaders, scientists, thinkers, educators and pioneers spent most of their time and energy learning in the inner realm before they came out with their new and creative ideas and projects. Of course, some will claim that their ideas and discoveries are their own but, the truth of the matter is that most if not all of our ideas come from or were rewritten or rediscovered from the inner libraries – including the contents of this book. Much of my learning, studies and research were done in or borrowed from the inner libraries of the Earth.

There are more places or states of consciousness that can be visited or experienced in the inner world than in the external world without limitations of transportation or time frames. What can limit people at the dream level or during out of body trips is their belief system or their mental and emotional programs. This explains why when people are in a dream state (emotional or mental body traveling), they think they cannot pass through doors or thick walls or that they need to drive to go somewhere else – because of the concrete mental program that we are only our physical body. Once you realize that you do not just exist on the physical level, but that you have energy bodies that are more refined, you can pass through walls or buildings in your dreams. You can go to another place just with intention, which allows you to transport yourself instantaneously in your dream state.

In the movie, "What Dreams May Come", Robin Williams is partially conscious of the inner world after his physical death. He has to learn how to fly, and by sheer intention, he is able to do it after changing his beliefs. The film industry has been generous to the public with films that roughly explain the inner world experience like that of "Sixth Sense", starring Bruce Willis. While existing in the inner world after dying, he is unaware at first that he is dead because he is the same emotional and mental being he was when he was alive. He only realizes he is dead later through another person. When people physically die, they are still the same person, with the same behavior and emotional and mental software – except for the dropping of their physical and etheric bodies.

When you sleep, you can meet or communicate with or even educate or counsel people who have died. You can find people in the inner world when you know that it's possible. Inner healing and recovery of unfinished business or relationships can be achieved during your out of body time, during sleep. In fact, some students of the Ageless Wisdom are trained to help pacify or heal people who have died as in incidents like the Twin Towers destruction in which many people died. Many of these healers went to help in their sleep, to console and heal the people who died. This is important because when people die through traumatic death, they have emotional and mental difficulty in the inner world. Therefore, this process can be a great service to many human beings anytime.

The Ageless Wisdom, particularly the teachings of Holy Master DK through Alice Bailey's books, mentions that there are different learning centers or states of consciousness in the inner world:

- The learning center for normal people – for ordinary subjects and topics
- The center for knowledge – for more intelligent human beings who want to specialize in different technicalities and bodies of knowledge.

- The enter of wisdom – for more advanced human beings who want to transmute knowledge into wisdom, including esoteric sciences or higher truth. This center also includes schools for Sainthood and the starting curriculum for Sages.

Very advanced initiates or the beings we call Sages need higher learning and a curriculum for higher evolution, which does not require sleep at their level. They can access the data or information through advanced meditation or out of body trips, but not necessarily to the emotional and mental world. They go to the world of higher beings – to the higher libraries for courses in higher evolution. This is the difference between super-advanced human beings' and ordinary people's process of enlightenment, experience and behavior. Their point of reference for truth and sources of information are already at a different level.

Through the conscious use of sleep and dream time – where the emotional and mental bodies can exit and go to an accelerated learning experience – not only will humans acquire knowledge but also wisdom. With this synchronized and parallel time to learn in sleep and while awake, the educational system will change, the learning speed and time will be shortened and the curriculum will definitely be upgraded into a super-advanced stage that synthesizes the lessons of both objective and subjective life.

There will be conscious continuity of the physical, emotional, mental and spiritual world, and these new realizations of continuity of consciousness will remove the gap between physical life and life after death – and will give a new meaning to immortality.

The right understanding of death, the process of discarnation and incarnation, will be seen as a natural process of evolution. If we do not see and appreciate it now, our future civilization will know, understand and celebrate it as a way of life.

Psychic powers

Tenth power: Teleportation

The psychic powers of teleportation are quite magical to the minds of people. Teleportation remains a fairytale in the minds of most people, even though it is portrayed frequently in the movies. Though I have not witnessed human beings teleporting in the commonly understood sense – where one physically move instantaneously from one place to another – we can briefly discuss this inner power and its potential applications.

Teleportation – as levitation

I have witnessed several people doing this. In fact, some yoga groups specialize in *mantras* or recitation of magical words that allow them to levitate or jerk up and down a few inches during meditation – after many years of practice. Some of my students have witnessed this display of inner powers.

For instance, while I was healing one of my clients in New York City, who is a psychic, she levitated more than 15 times. She was sitting on her bed in a lotus position, and after 10 minutes of energy healing work on her cancer, her whole body jumped up and down to a height of 18 inches while maintaining the lotus position – with no effort on her part. In fact, she did not bother to react because she was in a semi-trance and being a clairvoyant, it was kind of normal to experience something like this. When I finished healing her, she felt so good and relieved of her pain that she accepted all of it as part of her healing process. I observed that there was a big energy being released throughout her system, especially through the spine from below the perineum. It was like an energy pressure pushing her whole system up and causing the levitation of the body. This invisible energy and force is what we have referred to earlier as the sacred fire or the *kundalini*, which I will explain further in the next section on Spiritual Powers.

Physical teleportation

This is an unusual psychic power, but I will give a little background on it. I am sure this can be a future skill to be developed in humanity in the distant future when we will have good and safe reasons to utilize this power with more advantages than disadvantages.

During my trip to the Himalayas, I heard many stories from yogis and gurus that some advanced spiritual teachers can disappear and appear in different places instantaneously. I have not seen this myself. In the Philippines, there are also many stories about special teachers who are able to run on top of the tall grasses without touching the ground. I have not seen it myself, so I cannot confirm it. My sense, though, is that it might be possible because since there are laws of gravity, there must also be inner laws of anti-gravity producing levitation and teleportation.

This levitation is already being done partially by astronauts without effort in simulated zero gravity space, so it's about time for ordinary human beings to do it, even in normal Earth settings. Scientists are excellent advocates of replicating this possibility through technology. If the mind can conceive of the idea, the physical skill can be developed, and the reality will follow what the mind or imagination has conceived. If you have watched the movie "Crouching Tiger, Hidden Dragon", the actors were able to teleport and even fight while floating in the air. What's in the movies now might be done in reality soon. Science fiction movies forecasted the first man on the moon and the use of cellular phones. More and more movies are projecting this levitation possibility. Humanity has that potential in its seed of consciousness, so I believe it will happen in the future.

Teleportation through consciousness

The more advanced beings need not move the physical body to enjoy levitation or inner world trips. Their energetic and spiritual consciousness can teleport anywhere they need to go. They can transfer their consciousness not only horizontally within one plane, but they can also be consciously and simultaneously present in multi-planes of existences from the Earthly existence up to the spiritual world. This is the real inner power, and it will take longer time for the majority of humanity to be able to perform this.

Many initiates like advanced Sages are capable of teleporting through consciousness, not only in the planetary realm, but also in inner cosmic consciousness. This is the power that makes them different from ordinary Saints and much more different from ordinary humanity. But this is the goal for human beings – to be freer and to be liberated from the lure and pleasures of mundane living, not only traveling from place to place, city to city, one country to another, climbing a tall mountain from the foot to the summit, but also spiritual ascent to higher consciousness. Most of humanity's activities are still in the physical plane – the consciousness is still in the physical realm.

What about other worlds beyond your imagination and the imagination of our ancestors? There must be a greater adventure of life and experiences to explore and to expand to. We can only do this if we become more detached and free form the personality limitations of our flesh, lower passions and mental beliefs – and start opening our horizons of subjective reality and the aspiration to explore and master the path of inner power development through psychic powers. Advanced meditation is one of the important tools, plus the application of inner powers to the service of humanity and the world. Once spiritual development is a priority, all of these inner powers will bring you to the greatest experiences you can possibly imagine. It can be achieved, because by going through this book, you are already on your way up.

Section E
Spiritual Powers: Principles & Techniques

- **I. Questions about spiritual powers**.. page 290
  - A. Why the need for spiritual inner powers?........................ page 290
    - 1. Comparative study of levels of human development and inner powers & faculties.................. page 291
  - B. Why the need for enlightenment?.. page 295
    - 1. Comparative study of levels of human development and responsibility & service.................. page 296
  - C. Who am I?..page 297
    - 1. The triplicity of a human being..page 297
    - 2. Causal body development corresponding to stages of spiritual development................................ page 301
    - 3. Different stages of the development of lower bodies in humans..page 305
  - D. What can I be?.. page 307
    - 1. The path of human development & evolution............. page 307
    - 2. Stages of the spiritual path.. page 308
  - E. What are the major sources of spiritual powers & how can we access them properly & in the right dose & quality?..page 309
    - 1. Major sources of spiritual powers.................................page 309
      - a. Earth as a macrocosm... page 309
      - b. Earth's Higher Beings & groups..............................Page 316
      - c. The *Kundalini* or sacred fire.....................................Page 320
        - • Major functions of the *kundalini*......................page 322
        - • Practice 1: Nostril-controlled breathing technique to improve internal energy & balance the *kundalini*....................................... page 326
        - • Technical requirements for awakening the powers of the *kundalini*..............................page 329
        - • Methods to develop spiritual powers through safe awakening of the *kundalini*........page 331
          - - Physical exercise & breathing techniques..page 332
          - - Simplified *mantric yoga*............................ page 333

d. Power of virtues through culturing the personality.. page 338
- Benevolence...Page 339
    - Steps to develop & enhance benevolence...page 340
    - Technique to build goodwill........................ page 341
    - Releasing the negative past technique..... page 343
    - Recommended practices to develop benevolence................................ page 348
- Altruism..page 349
    - Techniques for practicing the habit of altruism.. page 352
    - Techniques for already generous people..page 354
    - Levels of altruism.......................................page 357
    - Things to avoid in order to unblock the flow of prosperity................... page 359
- Discipline & constancy....................................page 359
    - Techniques for building the virtue of discipline & constancy................. page 360
    - Energy centers that enhance the virtue of discipline & constancy ....page 364
    - Esoteric method to improve the virtue of discipline & constancy.. page 364
    - Objectivity & practicality.............................page 365
    - Methods to master the perception & validation of truth...................................page 366
    - Guidelines for digesting information & expressing practicality............................ page 368
    - Methods to apply discipline & practicality..page 371
- Will-power & vitality...page 374
    - Level of will-power development and the chakras & energy body(ies) employed... page 375
    - Techniques to improve will-power.............page 376

- Group consciousness................................ page 380
  - Requirements for group consciousness..................................... page 381
- Good health... page 386
  - Recommended practices for etheric & vitality health....................... page 389
  - Techniques to maintain emotional & mental health................. page 390
  - Techniques to maintain spiritual health a way of life.............................. page 391
- Virtue of sacrifice....................................... page 392
  - Techniques to intelligently apply the virtue of sacrifice.......................... page 393
  - MDP Inner Renewal Meditation......... page 395
  - MDP Soul Radiatory Blessings: Applications & Service........................ page 401
- Reminders about inner powers.................page 405

Section E
Spiritual Powers: Principles & Techniques

The subject of spirituality is as extensive as the study of cosmic life. Therefore, the teachings in this book are grounded in your immediate spiritual development needs and are designed to enhance your inner powers.

We will start studying from the generals to particulars, from universal principles to processes, which are broken down into methods, techniques and the required results in your daily life. In dealing with inner powers, we will lay a foundation to use them for maximizing your performance and balancing your life.

In the study and application of spiritual powers, it is best to start by asking the following five questions:

1. What is the need for spiritual inner powers?
2. Why the need for enlightenment?
3. Who am I?
4. What can I be?
5. What are the sources of spiritual powers, and how can we access the right dose and quality?

Let's discuss the questions and their answers individually, even though they are all related to each other.

First question: What is the need for spiritual inner powers?

The Earth and humanity are subjected to universal laws of evolution, and this is one thing that neither can resist nor escape. A cosmic intelligent force and energy stimulates, guides and fosters this growth, both on material and spiritual levels. Therefore inner power development is natural part of evolution, and it is a matter of time before one is stimulated with this impersonal Macrocosmic Power.

But why wait passively for the power to come in the far future? Why not cultivate it now, plant the seeds and harvest them soon? Saints and Sages harness wisdom through inner powers, which set them apart from normal philosophers or academics. Inner powers drive them to contribute more to the world than most people, but not necessarily in the material sense. The more advanced you are, the more equipped your life with inner powers.

To give you a concrete comparison of the relationship of the level of development in beings and inner powers applied, here is a chart for your study:

Level of Development	Description of the faculties & inner powers developed
Early Humans	• Sex instinct, instinct to survive through food & shelter • No health consciousness • Living only at the physical level
Average Humans	• Sex instinct, desire for a better home, food and passionate relationships • Living & conscious on physical & emotional levels • Power of lower desire & passion • Power of common sense & stimulus to learn up to the college level • Spirituality through religion
Intelligentsia	• Desire for a comfortable lifestyle with a bigger home, good food, recreation & more money

	• More conscious of physical, emotional & mental health • Desire for sex & good emotional relationships, expressed through family • Power of passion • Power of emotional intelligence (mystical type) • Power of the higher grade of common sense & concrete mind (mental type) • Aspiration to be educated up to higher levels or upgrading oneself through continuous self-education • Practical skills through seminars • Spirituality through religion & the more advanced employ practical spiritual training & meditation • Mystical types have psychic powers • Power for innovation • Initial power to create ideas or thoughtforms and organize them
Advanced Humans	• Aspiration for a cultured lifestyle with priorities for intelligent hard work guided by a great vision, mission & plan • The mystical type has the power of the heart & emotional intelligence • The mystical type has developed several psychic powers & the initial stage of intuitive perception • May or may not have the power of expertise

	or leadership • Partial integration of the concrete & abstract minds for mental type • Practical mysticism or combination of religions & other spiritual practices or services
Saints & Transfigured Masters	• Power to create & organize strong thoughtforms and ideas • Power of materialization of big ideas • Aspiration to live a life of high integrity, selflessness, altruism, benevolence & world service • Power of love & emotional intelligence with intuitive perception • Power of concrete & abstract mind integration, especially mental type • Power of spiritual persistence • Power of intelligent hard work • Power of visionary & servant leadership • Power of inspirational leadership by great example • Power of energy presence & magnetism • Starting power of the Third Eye
Sages	• Power of righteousness brought by their integration of the faculties of spiritual light, love and will-power • Power of spiritual freedom by finishing human karma • Power of love & wisdom

	• Power of detached compassion • Power of discrimination • Power of good intuition & the Third Eye • Power of spiritual oneness • Power to create & materialize thoughtforms and ideas with global influence • Power spiritual leadership up to global levels • Power to transform and lead groups of spiritual students & servers globally

As we study the different levels of development of human beings – their attitudes and aptitudes expressed either as faculties or inner powers – we can conclude that external powers, inner powers and people's contribution to Earth's and human evolution are all related. The amount of power you are endowed with is the power you can apply through your inner development. What makes people more successful and fulfilled than others is the ability to instinctively rise on every occasion to help others, solve problems properly, make good decisions and serve a great cause. This is a necessary attitude and aptitude, accompanied by the sense of urgency to accomplish goals and results. To be able to stay ahead and on top of things during this fast-paced modern life, the development of advanced faculties and new spiritual acumen is a must. Those who will respond to this new requirement will be the leaders of global leaders.

> People naturally follow, respect and trust leaders who are more powerful and spiritual than they are.

Second question: Why the need for enlightenment?

Enlightenment is not necessarily only a result of the practice of one's religion or merely through a life of prayer or meditation. I also believe that faith without work or action is not the best option. I frequently meet many business and political leaders who are more enlightened than prayerful people or meditators. Enlightenment is not merely a byproduct of sitting in a yogic position for a few hours a day, hoping that one day soon, a big bang explosion of light will happen. The process of growth towards greater light is a gradual undertaking that involves:

- Growth and maturity of the incarnated personality to use the body, emotions and mind properly to contribute to life, from a smaller scale to a global influence. This influence is directly proportional to spiritual and inner power development.
- How to make better and wiser decisions affecting not only the concerns of your family or group, but also employing more advanced faculties for projects of a larger scale with greater influence and service
- How to solve problems properly with fairness and inclusiveness, resulting in win-win situations
- Visionary leadership and the ability to materialize far-reaching goals effectively
- The ability to grasp larger schemes of global plans and cooperate with your chosen part in those plans
- Faculty of discernment and wisdom that leads to righteousness in all your ways

Therefore, if we compare these above requirements with the feats and inner powers of the prophets, Saints and Sages like Moses, Buddha, Jesus the Christ, Mohammad, Lao Tzu, Confucius and other founders of great world religions and movements, it follows that enlightenment is a practical power that can be used daily. Many great business leaders, politicians, educators, artists and scientific executives are also the modern embodiments of enlightened humanity. Many of them are great decision-makers and problem-solvers and implement services and projects that affect the world. If they can combine their material success and emotional and mental aptitudes with even higher training to develop their latent spiritual inner powers, some can be the future Sages in this century.

Let's analyze the following tabulation of human levels of development and the responsibilities and service rendered for each level:

Level of Development	Responsibilities & Service
Early Humans	To oneself & immediate family
Average Humans	To oneself, immediate family, relatives & friends
Intelligentsia	To oneself, immediate family, relatives, friends, colleagues, immediate communities & social affiliations
Advanced Humans	To oneself, families, relatives, friends, social affiliations, national or international projects & missions
Saints & Transfigured Masters	To humanity To environment & nature To other lower kingdoms To Earth as a home To other world groups working for Earth's evolution

Enlightenment, therefore, is a progressive growth towards the development of inner faculties and powers expressed through a higher cultured service and greater usefulness in the world. To the enlightened person, the outer life is a manifestation of the inner development, therefore you can determine how enlightened a person is by evaluating their degree of service and how much responsibility they handle.

Third question: Who am I?

> Religions and great philosophical teachings have always advocated the old adage: "Man, know thyself."

This is to keep human beings searching for truth about life and its correspondence to the Source of Life. If we believe in the spiritual teachings that humans are made in the likeness and image of the Creator, then by understanding ourselves, we have a corresponding idea of the nature of the Macrocosm, what people call God.

Taking this concept into a concrete study, we refer to the wisdom of the past ancient teachings of the enlightened teachers and Sages as the Ageless Wisdom. These ancient teachings have always advocated the idea that we are composed of a triplicity: the Spirit, the Soul and the incarnated human being called the personality. To lay out this basic truth, let's study the following:

Triplicity in a human being

1. **First Part of the Human Triplicity:**

   The Divine Self, The God within, the Spirit

   - This is a seed of Godliness in every human being, which is a fraction of the Divine Macrocosm. It is indeed an essence connected to and centered in the Supreme Divine Life on Earth. It serves as the basic seed or germ of divinity that sustains the will to live and evolve, containing the purpose of planetary life. Let us call this the Divine Human Spirit, or in Sanskrit, the ***Monad***, the **Divine Self**, the "**I am that I am**".

- The Divine Self or *Monad* is the one God within, but qualified with 3 divine qualities, which are a replica of the planetary deity:
    - Will and power: the first quality, first person, the Father
    - Love-wisdom: the second quality, second person, the Son
    - Active intelligence: the third quality, third person, the Holy Spirit

  At first this center of Divine life in a person is latent and unmanifested until the Soul and incarnated personality have evolved to merit from and contribute to its functionality.

2. Second Part of the Human Triplicity:

The Soul (The Higher Soul & the Soul's Causal Body)

- The **Higher Soul** is a being with consciousness that mirrors the qualities of will-power, love-wisdom and active intelligence from the Divine Self or *Monad*. The trinity of the Higher Soul is reflected from the trinity of the Divine Self.
- The Higher Soul of humans up to the level of a Saint is the transformer and transmitter of Divine electricity, Spiritual Power and Life from the Divine Self to the incarnated personality. At the Sage level and higher, the incarnated personality is already fused with the Soul and this new set-up is directly aligning to the Spirit or *Monad*

- The Higher Soul is the transmitter of experience and lower essence from the incarnated personality to the Divine Self. After a long series of lives as an incarnated human being, the accumulated substances and essence of experience are absorbed and assimilated by the Higher Soul, which will eventually feed the development of the Divine Self or *Monad*.

- The Higher Self is composed of 3 energy centers:

  - The *Atma* or spiritual will power center
  - The *Buddhic* or love-wisdom and intuitional center
  - The abstract mind center

  The 3 energy centers act like the transmitter and transformer of the 3 types of energies and forces embodied in the Spiritual Trinity of the Soul. Eventually, as the incarnated personality grows and graduates to the Sage level, these Soul centers will regulate the physical, emotional and mental bodies and consciousness if the human being is still physically incarnated.

- There is a Divine life and electricity supplied from the Divine Self to the Higher Soul and also from the Higher Soul to the incarnated person. This supply of power and life passes through a passage called the ***sutratma*** in Sanskrit, or let's just call it a **Divine Electrical Cord**.

- As the human being grows spiritually, there is a constructed cable of light called the ***antahkarana*** in Sanskrit that serves as a connector of consciousness and data from the incarnated person to the Higher Soul and vice-versa. The more advanced humans, Saints and especially Sages are equipped with a more developed "cable of light", allowing a free flow of tremendous information from the Higher Soul to the incarnated person and alchemized substances and the essence of experience from the incarnated person to the Higher Soul. This will only start to happen during the more advanced development of the person. It requires eons of time and many series of incarnations to be able to reach this stage where there is conscious contact between the two parties. Evolutionary progress and inner development are very fast after this connecting cable of light is constructed. Intuition and advanced mental faculties are available after this stage.

- The **Causal Body** is a part of the Soul that accumulates the substance and essence of experience from a person's many lives. The byproduct of each incarnation that brings consciousness is deposited in the Causal Body. It is like a sac containing the substance of experience and is made up of varying colors and densities of light. When this is substantially developed, there is an automatic brightening of the different bodies, especially the abstract mental body. The spiritual development of a person is directly proportional to the amount and intensity of light and substance in the Causal Body. That is why there is a concrete description of advanced humans as old Souls and average human beings as young Souls. This refers to the level of development of the Soul as manifested in the Causal Body's emptiness or fullness of the substance of experience.

Here is a rough estimate of the current standard of causal body development corresponding to different stages of spiritual development:

Causal Body Development Corresponding to Stages of Spiritual Development	
Early humans	Empty to a small percentage of substance
Average human beings	Less than 20% full & more than 10%
Intelligentsia	At least 20% full
Advanced human beings	At least 35% full
Saints	**Mystical**: Approximately 40% full **Esoterically trained**: At least 50% full
Sages	Full causal body & substances being completely diffused to the Higher Soul
Super Sages	No need for causal body

3. Third Part of the Human Triplicity:

Incarnated Person (The Egoic Lotus & the lower bodies)

- **Egoic Lotus:** The incarnated personality is a byproduct of the Higher Soul extending itself to the lower planes of the Earth via the Egoic Lotus. The Higher Soul existing in a higher state and frequency needs lower energy vehicles and physical bodies to experience the life of the mental, emotional and physical planes of the Earth.

The Egoic Lotus is currently shaped like a flower with 3 tiers of petals for the more advanced human up to Saint level, but it looks like a closed bud for the intelligentsia level. When it's open, it has a total of 18 petals categorized as follows:

- 3 petals of will-power: connect to the physical and etheric body
- 6 petals of love-wisdom: connect to the emotional body
- 9 petals of active intelligence: connect to the concrete mental body

There is also a jewel inside the Egoic Lotus serving as the spiritual software for the incarnated personality. This gradually manifests as the incarnated personality evolves to a more advanced stage.

The Higher Soul uses the Egoic Lotus as the controlling mechanism and software of the incarnated personality up to the Sage's level, after which the Egoic Lotus and the Causal Body are no longer needed. and the accumulated substance in the Causal Body is diffused to the Higher Soul. At that time, the Higher Soul energy centers will control the lower bodies directly if the person remains incarnated. The Egoic Lotus also serves as a transformer and transmitter of Divine electricity and information from the Higher Soul to the incarnated bodies. In addition, it acts as a converter and transmitter of the incarnated personality's substance of experience and alchemized energy for the Causal Body to deposit. In more advanced humans up to the Saint level, the Egoic Lotus is an orifice to construct the *antahkarana* or the connecting cable of light or path of consciousness for the Higher Soul and the incarnated personality.

- **Lower bodies:** What most humans know as the real person is just the physical body. The intelligentsia start to be familiar with the functions of the various different bodies, namely:

- physical and etheric body for physical movement and vitality
- emotional or astral body for feelings and emotions
- concrete mental body for thinking and reasoning

Again, this triple manifestation of the incarnated person is a mirrored aspect of the Divine Self and the Higher Soul. As the human being develops to a higher level, the Higher Soul regulates and empowers the incarnated person directly as follows:

- *Atma* or Soul's will-power over the physical and etheric body
- *Buddhic* or Soul's love-wisdom over the emotional body
- Abstract mind over the concrete mind

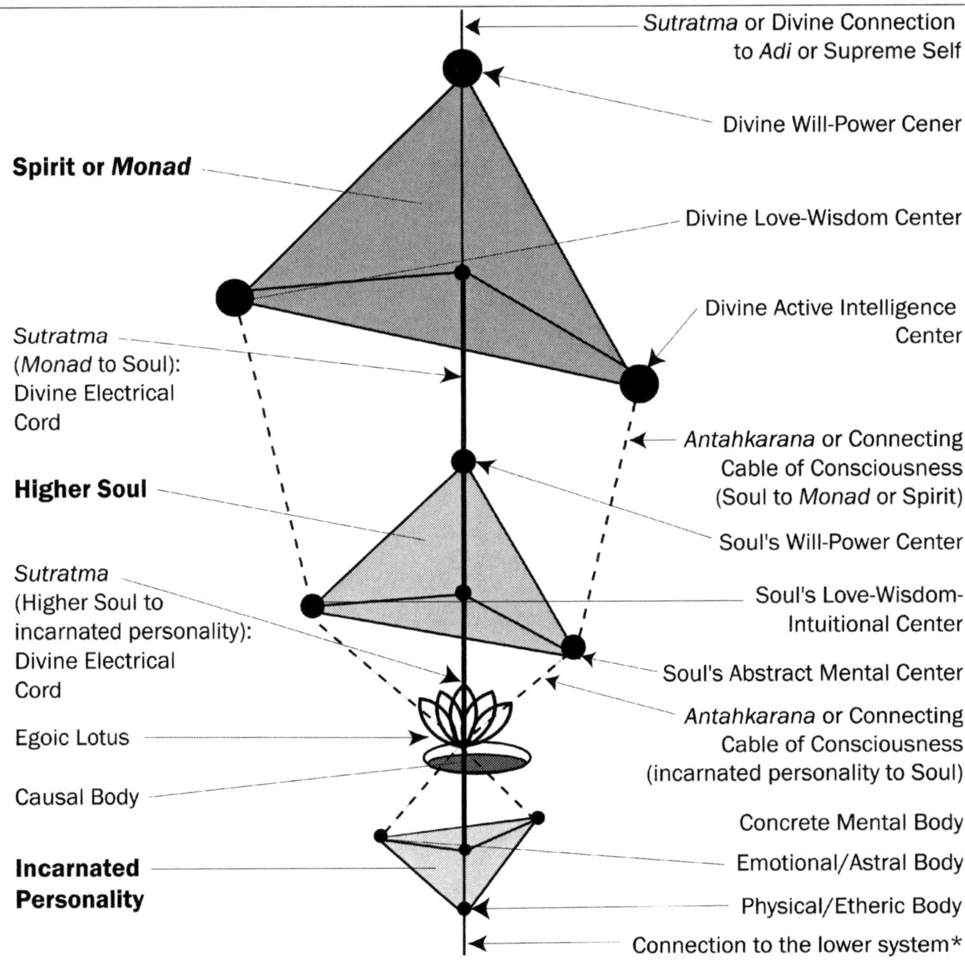

Schematic Diagram of the Human Triplicity: Spirit, Soul and Incarnated Personality
Figure 31

Evolutionary Path:

- **Path of Divinity**: Evolution from the Higher Soul to the Spirit (concerning Sages up to Super Beings)
- **Path of Spirituality**: Evolution from the incarnated personality to Higher Soul development (concerning human up to Saint's level)

* **Lower system**: This is only available at a very advanced level of human development wherein the evolution of the Egoic Lotus, Soul and *Monad* have lower correspondences manifested already. The lower system requires a very high degree of *kundalini* awakening to unfold. This is discussed in detail in Volume II of *Inner Powers*.

The following list will show you the development stages of the different bodies in humanity:

Different Stages of Lower Bodies Development in Humanity	
Early humans	Physical & etheric integrating and the emotional body is starting to be stimulated. The mind is dormant.
Average humans	Physical & etheric body integrated, emotional body is active and concrete mind is starting to be stimulated. The abstract mind is dormant.
Intelligentsia	Physical & etheric body & emotional body are starting to integrate, concrete mental body is active & abstract mind is starting to be stimulated
Advanced humans	Physical, etheric & emotional bodies are integrated, concrete & abstract mind are active & starting to integrate
Saints or Transfigured Masters	Physical, etheric, emotional & mental bodies are integrated.

"Integration" in the above cahrts refers to additional functionality and admixture of usefulness rather than fusion of the energy bodies.

Therefore, the abstract question, **"Who am I?"**, can be perceived concretely and summarized as follows:

- **A human is a being with a Divine Self or Spirit utilizing the Soul to express life in the Earth's lower planes via the many series of**

lives of the incarnated personality. The purpose of the Divine Self investing its Divine life and electricity into the Soul and the incarnated personality is to gather experience and bring it back as consciousness and substance of enriched life. It is like saying that the investor, the Divine Self, will gain its capital plus interest or profits. The Divine Self, after the purpose is achieved, will also evolve and grow into a more mature being, which then becomes a brighter filament of light on the higher planes of Earth's existence. The physical, etheric, emotional and mental bodies are temporary sheaths and vehicles for the Soul to control the personality. These bodies change in each incarnation to suit the Soul's goals until the entire Soul's purpose is achieved. By that time, the Soul chooses to withdraw completely from the Earth's lower planes or comes back as another enlightened personality for planned Earth service.

- **The human being is an entity mirroring the quality of the trinity of the Divine Creator, the Macrocosm**, through the aspects and qualities of:

  - Will-power
  - Love-wisdom
  - Active intelligence

  This explains the teachings of the Divine trinity and mysteries in many religions.

- **The experience and effort of the incarnated personality makes the Soul and the Divine Self evolve from dormancy to an active state**, **from an unmanifested state to a manifested dynamic state.** The life and activity of the incarnated person, therefore, is very important to stimulate the evolution of its real owner, the Soul – who is going to stimulate the development of its sleeping owner, the Divine

Self or *Monad*. The traditional concept of the old school that the lower bodies and the incarnated personality are not important is counterproductive. Since the lower self, the incarnated personality, plays a significant role in the evolution of the Spirit, it also plays a significant role in the transformation of the Earth.

Now, there is another point of interest in most searches for Truth.

Fourth question: What can I be?

The usual philosophical question, "where are we going?" is sometimes misleading because it tends to connote displacement of space and time rather than future development or expansion of consciousness. Thus, the question, "what can I be?" is better. The simplest way to answer this is based on our list of the different stages of human development. The "I" depends on your stage of development. Therefore we can infer the following:

The Path of Human Development & Evolution
Early human level persons become average humans
Average humans become intelligentsia
Intelligentsia become advanced humans
Advanced humans become Saints
Saints become Sages
Sages become Super Sages
Super Sages become Super Beings

Then these stages of evolution create the qualifications for the spiritual path as follows:

Stages of the Spiritual Path
1st Stage: Merely physical life existence satisfying physical appetites & requirements, like early humans
2nd Stage: Life of physical & emotional activities, like that of average human beings with mass consciousness and still spiritually asleep
3rd Stage: Life of physical, emotional & normal mental activity, like that of the intelligentsia, who are starting to be spiritually awake
4th Stage: Life of physical, emotional & mental activity partly inspired by a higher spiritual impulse from the Soul as exemplified by the more advanced human beings
5th Stage: An integrated life of the physical, emotional & mental activity with a great deal of spiritual purpose & guidance from the Soul as shown by the life of a Saint
6th Stage: An integrated life of the incarnated personality as completely infused by the Soul and starting to be guided by the Divine Self – a Sage's life
7th Stage: An integrated Soul life guided by the Divine Self – the Super Sage
8th Stage: A complete life at the level of the Divine Self or Spirit, integrating unto itself the life of the Soul as in the case of Super Beings

In summary, the future of the incarnated person is to become the living Soul, and the Soul's future is to be a living Spirit or Divine Self. This is the simplest, most succinct way of expressing the goal of the spiritual path and what we can be. To evolve to the next stage of development and to embody that stage and path is the spiritual strategy. All the activities and careers of each life are but personal ingredients that serve as a curriculum for each grade of achievement…believe it or not.

Fifth question: What are the major sources of spiritual powers, and how can we channel the right dose & quality?

Major Sources of Spiritual Powers
Earth as a Macrocosm
Earth's Higher Beings and Groups
The *Kundalini* or Sacred Fire
Virtues and Culturing the Personality

There are 4 major different sources of inner power for a human being. We'll discuss each of these in detail.

Major sources of spiritual powers

First major source: Earth as a macrocosm

There are different sources of spiritual powers from the Earth as an alive Being. The Earth, serving as the Macrocosm of some higher spiritual beings, human beings and all kingdoms of nature, provides varying levels of experiences, spiritual energies and substances that empower every being, visible or invisible, higher or lower. Even the Super Beings, Super Sages, Sages and Saints are still inside the planetary Earth consciousness. Their residence is at the range of frequencies and Earth planes where their energy bodies are located and vibrating. The Spirit and Soul of human beings are also accommodated in different levels of the Earth planes or dimensions. Therefore, the main sources of inner powers of the Earth's population and occupants of the Earth as a Macrocosm are as follows:

Sources of Inner Powers from Earth as a Macrocosm
Supreme Earth plane or *adi*
Monadic or Divine Spirit Earth plane
Atmic Earth plane
Buddhic Earth plane
Abstract mental Earth plane
Concrete mental Earth plane
Emotional Earth plane
Physical & etheric Earth plane

The above table lists several planes or dimensions of the Earth that are like energy worlds within themselves, thus the more refined levels interpenetrate the other less refined dimensions.

Different planes of the Earth as a macrocosm

- **First highest plane: Supreme Earth plane**

  This is called the *Adi* in Sanskrit or highest Divine plane composed of seven layers of frequencies. The teachings of the Ageless Wisdom consider this plane as part of the most refined or first etheric plane of the Cosmic Being's aura. Therefore, the Super Beings functioning at the *Adi* energy center wield the power of this plane, already employing a cosmic quality. This is loosely called "the seventh heaven."

- **Second plane: *Monadic* or Divine Spirit Earth plane**

  This plane is where the *Monad* or the Divine Self of human beings are housed. The *Monadic* plane is part of the second etheric frequency of the Cosmic energy body and is also called "the sixth heaven". It is made up of seven layers of frequencies. Those whose *Monadic* energy centers are active and functional get their energy and power from this plane.

- **Third plane: *Atmic* Earth plane**

  The *Atmic* plane layers of frequencies and is considered part of the third etheric frequency of the Cosmic energy body. It is called "the fifth heaven", and it houses the *atmic* or spiritual will-power centers of human Souls. The inner power in this plane stimulates will-power, fearlessness, detachment and spiritual centeredness. In Sages, this plane is the source of all of the abovementioned powers and qualities.

- **Fourth plane: *Buddhic* Earth plane**

  This plane also has seven layers of frequency and is called "the fourth heaven". It is part of the fourth etheric frequency of the Cosmic energy body. The inner power at this Earth plane includes universal love, wisdom and intuitional faculties. This is the first plane in which a being becomes really universal because it is the first layer of the cosmic aura. The Soul's *Buddhic* or love-wisdom center is housed in this plane.

- **Fifth plane (upper frequencies): Abstract mental Earth plane**

  This plane is called "the third heaven" and is composed of three layers or frequencies of abstract mental substance. It is the start of the spiritual plane of the Earth, housing the abstract mental energy centers of human Souls. The powers available in this plane are the abstract ideas and seeds of creative concepts with universal applications. This is where philosophers and abstract thinkers get their new information and advanced principles.

- **Fifth Plane (lower frequencies): Concrete mental Earth plane**

  This is considered "the second heaven", and both the concrete and abstract mental planes are part of the gaseous aspect of the Cosmic being's body. This plane has four layers of frequencies of concrete mental substance. The power available in this plane is logical and sequential thinking and the power of reason. Many aggregate and accumulated thoughts of humanity are deposited in the concrete mental plane of the Earth: positive, negative, true and illusionary. This is the big issue for this plane because of the contamination of lower vibrations of thoughts from all kinds of sources, including the formulated thoughts of past civilizations, present humanity and polluting and negative programs from the involutionary forces. Therefore, the power tapped from this level deals with both good and bad sources and qualities.

6. **Sixth plane: Emotional Earth plane**

   This plane is part of the liquid aspect of the Cosmic Being's body. It is also called the astral plane whose three highest layers of frequency are loosely considered "the first heaven". The lower frequencies at about the second and third layers of the Earth emotional plane form the state of purgatory, and the lowest frequency houses the state of hell. This is the issue of the emotional plane; serves as the transitional stage for people who have died. When people at the early and average human levels physically die, their next residence is the purgatory level to process their emotional bodies. The members of the intelligentsia who have lived negatively also pass through these Earth emotional frequencies, and the really bad humans have to be processed in the Earth's incinerator, "the hell world", to be purified energetically. The more negative and contaminated the emotional body, the longer it needs the purifying process. Hell serves as a junkyard and garbage processing plant for the Earth.

When a person dies, he or she still has the emotional and mental body as the extension of the Soul. The person is the same except for the absence of the physical and etheric bodies. Therefore, as soon as the right frequency of the emotional body is attained after death, the person can go to the higher astral plane, to the first heaven. There is a point when the Soul discards its connection from the emotional body, thus, the emotional body dies. This is the second death of the person. The mental body is still intact, and the activity will subsist only at the concrete mental level of the Earth called the second heaven. After the concrete mental body is processed, the Soul also withdraws from the concrete mental body, and this also dies. This is the third death of the incarnated person. At this point, the human being resides completely at the Egoic Lotus until the next incarnation when the Soul creates its next mental, emotional, etheric and physical bodies. The powers derived from the higher frequencies of the emotional Earth plane are emotional love, courage and emotional devotion and passion.

- **Seventh plane: Physical and etheric Earth bodies**

These are the densest planes of the Earth. The etheric and physical planes are part of the solid aspect of the Cosmic Being's body. The etheric part of the Earth has four layers of frequencies: atomic or first etheric, sub-atomic or second etheric, third etheric and fourth etheric. The physical aspects of the Earth are the gaseous, liquids and solids.

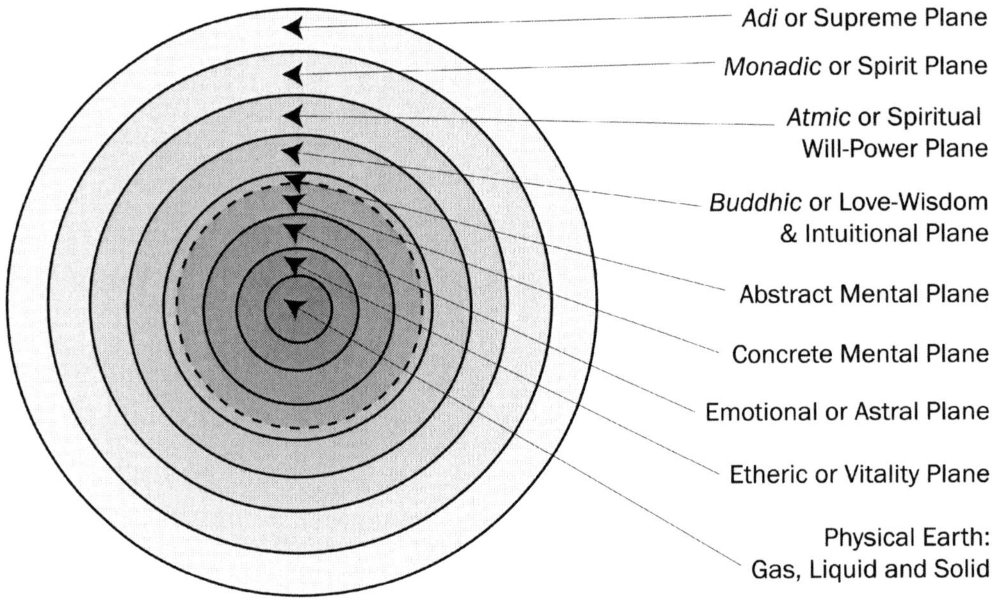

Earth Being as a Macrocosm:
7 Different Planes Composed of 49 Layers of Frequencies

Figure 32

Humanity has focused a great deal on the physical plane, using and also misusing the physical powers from minerals, metals, chemicals and gaseous substances. Humanity has used this power through alchemy, creating compounds and alloys and harnessing the power of the atoms to implement humanity's created goals. The etheric energies are used in vitality and stamina development, as well as in different healing modalities. Humanity has discovered and demonstrated the power of the body and stamina through sports, the Olympic Games and internal martial arts.

The different beings and kingdoms derive from or connect to different sources of powers and reside in different planes of the Earth as follows:

Sources of Spiritual Powers for Different Beings & Kingdoms	
Divine Beings & Super Beings	*Monadic*, *Adi* planes & beyond. The incarnated Super Beings with physical bodies have effects down to the physical planes.
Super Sages & Sages	*Monadic, Atmic, Buddhic* & abstract mental planes. The incarnated ones have influence down to the physical planes.
Saints	Partly *Buddhic* plane & mostly abstract mental plane & down to the physical if they are incarnated
Advanced human beings	Partly abstract mental plane & mostly concrete mental plane down to the physical plane
Intelligentsia	Concrete mental plane down to the physical plane
Average humans	Partly concrete mental plane & mostly emotional plane down to the physical plane
Early humans	Slightly emotional plane & mostly physical & etheric planes
Animals	The more advanced species & domesticated animals are starting to vibrate at the emotional plane & mostly at the physical & etheric planes. The normal animal species are more at the physical & etheric planes, minimally influencing the

	emotional plane.
Plant & vegetable kingdom	The more advanced types are starting to be affected by the lowest frequencies of the Earth's emotional body, with most activity at the etheric & physical levels. Most of the species of the plant & vegetable kingdoms are at the etheric & physical planes.
Mineral kingdom	The more advanced minerals are starting to vibrate at the etheric levels, especially the radioactive types. Most minerals are still at the physical levels.

The more subtle and refined the frequency tapped from the Earth planes, the higher the inner powers developed and utilized.

According to this understanding, radioactive elements are more advanced than the normal non-radioactive materials just as gamma rays are more potent than x-rays.

Major sources of spiritual powers

Second major source: Earth's Higher Beings and Groups

The Earth as an alive planetary being also has different groups as sources of inner power. Most of the more advanced groups are invisible and unknown to most humans. They are, however, slightly touched upon in the holy scriptures of religions. The writings of the Sages and their students in the Ageless Wisdom have indicated their existence. The books of the Super Sage, Djwhal Khul, as dictated telepathically to Alice Bailey is the most complete written body of information to date on these sources of planetary powers. The next two volumes of *Inner Power* will also describe them at greater length. Let me enumerate these sources of great powers on Earth:

Earth's Higher Beings and Groups
Shamballa
Extra-planetary Beings
Spiritual Hierarchy

- **Earth's Higher Beings and Groups:**

  First group: Shamballa

  This planetary center of power serves as the head chakras of the planetary being and is the main source of Divine energy and power on Earth. This power center vibrates at the *Monadic-Adi* planes and higher. It is the source and custodian of the will and purpose of the planetary life and the main entrance of Cosmic electricity and life force. It is also the main "satellite dish" receiving cosmic data for the Earth's existence as part of the Cosmic Being. As far as humanity and lower kingdoms are concerned, this is the source of absolute power on Earth. Members of the Shamballa center can be permanent, transitionary or temporary. The head of this Shamballic Center is a Divine Being called by the Ageless Wisdom, Lord of the World, S. Sanat Kumara. He/she is also described in the Bible as Ancient of Days and Youth of Eternal Summers. This Being is said to be the Great Planetary Earth Sacrifice who stays with this planet until its purpose is achieved. I will mention more details on this subject in the next two volumes of *Inner Powers*.

- **Earth's Higher Beings and Groups:**

  Second group: Extra-planetary Beings

  Several beings serve as intermediaries with the solar system or Cosmic Being for the needs of the Earth. They act as dispensers of different substances that the Shamballa, Spiritual Hierarchy (see below) and kingdoms of nature require. Peace, Mercy and other spiritual human needs are delivered through the Earth's higher officials, the Shamballa and the Spiritual Hierarchy from these Extra-planetary Beings. Other Cosmic Beings also forward their blessings and information directly to the Shamballa or the Spiritual Hierarchy.

- **Earth's Higher Beings and Groups**

  Third group: Spiritual Hierarchy

  This integrated group of illumined beings serves as the heart chakra of the planet. The group resides mostly in the *Buddhic* and *Atmic* planes, with the Saints residing partly at the abstract mental plane and the most advanced members residing partially up to the *Monadic* or *Adi* planes of the Earth. There are three major officers leading this planetary center and spiritual group:

  1. Lord Manu as the manager of will-power
  2. Lord Christ, also known as the Imam Mahdi, the Messiah and the Boddhisatava Maitreya, as the manager of love-wisdom
  3. Lord Mahachohan, the Lord of Civilization, as the manager of active intelligence

The Spiritual Hierarchy is composed of many departments that give information, guidance and protection and dispense power to all the kingdoms, especially humanity. All the brightest ideas and breakthroughs in human endeavor were and are inspired and precipitated by at least seven major departments:

7 Major Departments of the Spiritual Hierarchy	
Will-Power Department	Purification & destruction of obsolete forms and provides the force in the construction of new physical forms and pioneering works; politics & esoteric science
Love-Wisdom Department	Education, religion, psychology & healing
Active Intelligence Department	Business, entrepreneurship & philosophies
Department of Creativity & Arts	Music, arts & creative endeavors and efforts at harmonizing conflicts
Concrete Knowledge & Science Department	Concrete knowledge & sciences
Department of Abstract Ideals & Devotion	Churches & faiths, military and devotional & idealistic groups
Department of Ceremonial Order & Magic	Organizers of systems & forms, ceremonial magic & ritual groups

There are smaller subsidiary departments that work on many specializations of service and global projects, with some esoteric and many exoteric groupings. These spiritual groups are also broken down into many lineages headed on the physical plane by many spiritual teachers, politicians, scientists, educators, religious leaders, artists and business leaders. The real source of power for humanity is these abovementioned groups. Without their continuous stimulation and empowerment of humanity and the other kingdoms, the evolution of the world and humans would be very slow with no purpose or guidance. Therefore even the greatest of the Saints and world leaders have minimal power without these Divine and Spiritual groups. In short, the Souls and Divine Selves of humanity depend upon the guidance, teachings and blessings of the Shamballa and the Spiritual Hierarchy.

Major sources of spiritual powers

Third major source: The *Kundalini* or Sacred Fire

Whether the Shamballa group, Spiritual Hierarchy or human beings, the major fertilizer for inner power and spiritual development is the awakening of the *kundalini* or Sacred Fire. Even the Earth Being and the other Macrocosms depend upon the awakening of their *kundalini* for their inner powers.

A tree needs minerals and water to grow and bear flowers or fruit, or to at least stay alive and healthy. Super Beings, Sages and human beings also need spiritual minerals or spiritual food and spiritual water to grow and bear the fruit of enlightenment. The mineral or fertilizer is the *kundalini*, and the sexual energy is like the water that brings the *kundalini* to all energy centers and bodies. In ancient times, the *kundalini* was symbolized as the dragon spitting fire and the snake coiling up around a staff or cane. In India, some of the statues of their advanced yogis have cobras at the back of their heads. The number of snakes signifies how advanced the yogi or how many *kundalini* channels are opened. The founders of medicine symbolized its ancient wisdom with the caduceus – two snakes coiling around and ascending a staff with two wings and a round top.

Cobras as a Symbol of the *Kundalini*
Figure 33

Super Being Level with 49 Layers of *Kundalini* Awakening Symbolized by 7 Cobras at the Back

Super Sage Level with 35 Layers of *Kundalini* Awakening Symbolized by the 5 Cobras at the Back

Schematic Illustration of the *Kundalini* and the Medical Caduceus
Figure 34

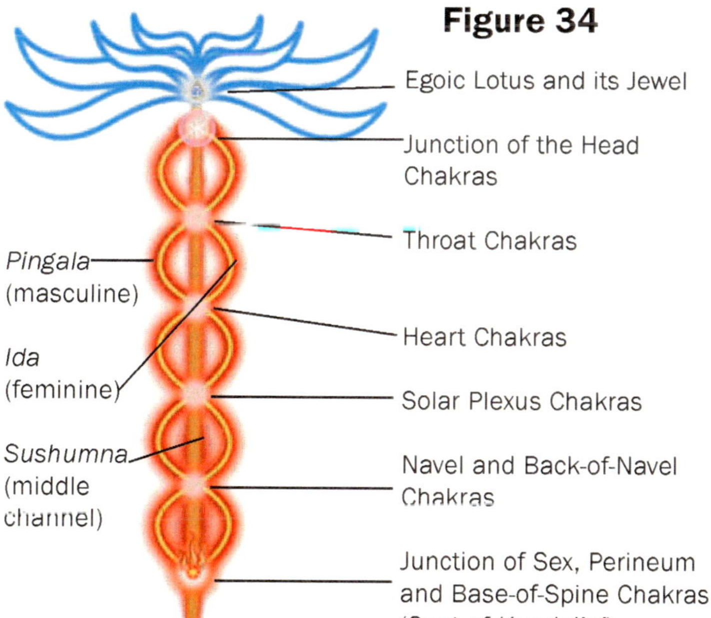

Medical Caduceus

As there are seven Earth planes with seven frequencies, each plane totaling 49 frequencies, the Super Beings at higher evolution within the Earth's requirement have to reach a total of 49 degrees or layers of *kundalini* awakening. Every human being has a certain level of *kundalini* development. It is a matter of the degree of opening of the sheaths covering the *kundalini* path. The real source of stimulation of the inner powers at all levels is how many of the *kundalini* layers are opened, to what level of energy centers and energy body of the incarnated person, Soul or *Monad* the *kundalini* has reached. The greater the opening, the higher the development of the being and the more advanced the power. A sample list of *kundalini* awakenings and stages of development with complete illustrations is available in Volume II of *Inner Powers*.

All the faculties, inner powers, energy levels, sizes of the charkas, development of the energy bodies and the degree of world influence of any person or being is directly proportional to their *kundalini* awakening. Here are some of the major functions of the *kundalini*:

Major Functions of the *Kundalini*
A catalyst & fertilizer to stimulate and activate the growth of the chakras & energy bodies
Serves to vivify the latent qualities & functions of the psychic & psychological powers, both positive & negative
Burns the protective webs or energy sieves separating the different bodies from physical up to the *Monadic* & *Adi* for the human being to be perfected
By the alchemy of the *kundalini* mixed with different energy & forces from the energy bodies & chakras, the resulting substance is brought to the Egoic Lotus at first, then to the Soul centers at a higher development

level in order to accelerate the Soul development. At a high stage like the Super Sage and Super Beings, the alchemized *kundalini* has to be brought to the *Monadic* level.

> This alchemized *kundalini* substance serves as a catalyst & fertilizer for the Soul to grow & for the different levels of Soul & *Monadic* development to manifest. This last stage of development applies to Saints, Sages and higher Beings.

The *kundalini*, resulting in the awakened development of the incarnated person, serves as a substance to fill up the *antahkarana* or spiritual cable connector. Thus, there is no true construction of this cable of consciousness without *kundalini* substance.

> Enlightenment & higher evolution are not possible without advanced *kundalini* development.

The *kundalini* path, when developed, acts as a rod or electrical conductor & transmitter of vast amounts of information & data. When perceived as a spiritual technological function, it serves as an optical passageway of tremendous data and information and at the same time as a power supply. It is super communication & electrical equipment when it is highly developed and activated.

The substance of the *kundalini* is the power that materializes thoughts, intentions & goals at a faster speed and greater potency.

Energy form and myth of the *kundalini*

If you study medicine's caduceus symbol deeply, you will see that it represents the inner path of the sacred fires latent in every human being. The central line is equivalent to the middle energy meridian passing through the central axis of your body. The intersecting snakes in the caduceus represent the masculine and feminine equivalent forces and the meridians traversing through the different roots of the chakras going upwards to the crown center. The wings on top of the caduceus are the equivalent symbol of the faculty of intuition and higher expansion when the *kundalini* is able to reach to the Egoic Lotus or higher. The round top of the caduceus is equivalent to the mental seed of consciousness seated at the pineal gland area inside the head. Once meditators understand this universal symbol, real yoga and spiritual practice can achieve greater heights of experience and inner power development. All ancient Sages and Enlightened Beings have used this knowledge to develop techniques to manipulate the sacred fires and develop the inner powers in all the energy bodies and centers.

The story of Prince Charming kissing and waking up Sleeping Beauty is a great fairy tale that hides the inner meaning of how to awaken your *kundalini*, sometimes called the sleeping serpent or dragon. Prince Charming is the Divine Light from the Soul, and Sleeping Beauty is the dormant energy (in most people at their current level of development) at an area slightly above the perineum, which moves upward with a spiral action. There are three major ascending paths of the *kundalini* fire or substance in the energy bodies:

1. The middle path or axis called the *sushumna* in yoga
2. Active force or the masculine aspect expressed as *pingala*. This is the path of the sacred fire that is affected by right nostril breathing.
3. Passive force or the feminine force expressed as *ida*. This is the path that is affected by left nostril breathing.

The advanced yogis and Sages have found that by balancing these two flowing forces or meridians (*ida* and *pingala*) through the balancing of the two breaths of the nostrils, the *kundalini* naturally flows upward.

Once the sacred fire of the *kundalini* circulates outward, it vivifies and fertilizes all the centers and their psychological as well as spiritual powers. This is what advanced meditators and yogis are technically doing to achieve great inner powers and advanced faculties. Even without formal schooling, their faculties of wisdom and higher qualities are very developed. It is not possible to divulge the advanced techniques of awakening the *kundalini* publicly in this volume, but a safer method will be given for practice. Let's first study the significance of the nostril breathings.

Importance of nostril breathing

According to ancient teachings in yoga, which I have experimented with extensively and have proven to be effective, if you breathe repeatedly only through one of your nostrils, there are certain outcomes such as:

- Right nostril breathing:
    - increases dynamic activity
    - increases sexual vitality
    - improves digestive functions
    - gives a very optimistic attitude

Slow and very passive people should apply the right nostril breathing daily for about 5 minutes. Sexually over-active or very aggressive people should minimize using this technique.

- Left nostril breathing:
    - improves passive and gentle nature
    - relaxes and calms you down
    - improves receptiveness and patience
    - gradually helps neutralize aggressiveness and superiority complex

In fact, you can observe that most healthy people have a natural alternation of breathing, either predominantly through the left or right nostril following what they call the circadian rhythm, which normally alters left and right nostril opening about every 90 minutes. The erectile tissues lining the internal nasal passage have this natural intelligence allowing alternate swelling to close and open each nostril. If the breathing of the two nostrils balances, you also feel a greater inner well-being. Here is the technique to safely activate the *kundalini* and allow it to flow naturally:

Practice #1: Nostril-controlled breathing technique to improve internal energy and awaken and balance the *kundalini*

(Practice for 10-15 minutes 3 times weekly)

Preparation

- Sit in a comfortable position with your spine straight and vertical and your feet flat on the floor. You may also sit in a lotus or semi-lotus position. Close your eyes for better concentration.
- Do 3 sets of the MDP Synchronized Breathing Technique.

Procedure

1. **Right nostril-controlled breathing**: Use your right middle finger and thumb to alternately close each nostril.
   a. Close your left nostril with your middle finger and breathe slowly and fully through your right nostril. Hold your breath for a few seconds.
   b. Simultaneously release your middle finger from the left nostril while closing your right nostril with the thumb.

Exhale slowly and completely only through the left nostril. Hold your breath for a few seconds.

c. Repeat steps a and b for 3-5 minutes only.

The goal is to inhale through the right nostril only and exhale through the left nostril only.

2. **Left nostril-controlled breathing**: Use your left middle finger and thumb to alternately block each nostril.

   a. Close your right nostril with your thumb and breathe slowly and fully through your left nostril. Hold your breath for a few seconds.

   b. Simultaneously release your thumb from the right nostril while closing your left nostril with your middle finger. Exhale slowly and completely only through the right nostril. Hold your breath for a few seconds.

   c. Repeat steps a and b for 3-5 minutes only.

   The goal is to inhale through the left nostril only and exhale through the right nostril only.

These techniques are not recommended for people with high blood pressure, heart ailments and cancer or for pregnant women. They tend to have side effects on them. If in doubt, consult your medical doctor.

You may choose to do only one type of nostril-controlled breathing to satisfy any physical or psychological requirements. If you want to balance your feminine and masculine internal energy, symbolized by the Chinese principle of *Yin* and *Yang*, you can spend equal time on both types of breathing or a little bit more on one type depending on what you want to decrease or increase in your functions. Refer to the discussion on the predefined outcomes of the two types of nostril-controlled breathing above. When these two breaths from both nostrils are balanced, the *kundalini* flows more smoothly and naturally.

What causes the awakening of this sacred fire?

When you inhale fully and hold your breath for a few seconds, the Divine light descends through the crown chakra and floods the bodies. When you exhale fully and hold for a few seconds, the *kundalini* is stimulated and increases its activity. By doing this alternately for 3-5 minutes for each nostril, there is already a substantial awakening of this magical fire. The outcome is increased internal energy in your bodies. This creative fire also vivifies and fortifies the chakras and bodies that it touches.

Of all the topics that engross esoteric schools and advanced schools of yoga, the *kundalini* has been a source of fascination and also a source of menace and fright to those who are ignorant of it. The true esoteric scientists, divine alchemists and white magicians are masters of the *kundalini* science. This power is the secret behind extraordinary feats in some yogis and spiritual teachers.

The right dose and quality of the *kundalini* awakening depends on one's level of spiritual development. It is very dangerous to awaken the *kundalini* power without proper guidelines and guidance from a knowledgeable teacher. It is like the strictness needed in handling the power of a nuclear bomb compared to handling only an air gun or dynamite. The greater the force and power, the greater the safety measures and knowledge required for its care and use. The technical requirements of awakening the powers of the *kundalini* are:

Technical Requirements for Awakening the Powers of the *Kundalini*
• Knowledge of the *kundalini*, its nature, function and mechanisms, including the do's and don'ts of handling it
• Purification of the energy bodies and energy centers and removal of energy obstacles along the front, back and middle meridians of the bodies
• Character-building to eradicate negative attitudes and behaviors, which get amplified by the awakening of the *kundalini*. The *kundalini* is like a fertilizer facilitating both the personal "weeds" and "flowers" to grow fast.
• Preparing the physical and energy bodies by proper diet and culturing of the physical body with special exercises and breathing techniques. People with vices like smoking, alcoholism, and substance abuse cannot safely practice the awakening of the *kundalini*.
• Outlets for the excess additional powers and energy as a result of the *kundalini* awakening for greater service and a bigger purpose in life. When the *kundalini* gets activated, tremendous power is added to your normal energy level. If this is not utilized for a good cause, it can result in congestion or over-stimulation affecting one's health. Also, additional energy and power is supposed to be sought for greater service or applications, not just for power hungriness. Good intention is important right from the start. Greater service guarantees that if your negative karma is accelerated as part of spiritual development and inner purification, the good karma generated from service can neutralize difficult situations.
• Activation and alignment of the chakras, especially certain

	important pairings and triangulations. This involves the three energy bodies.
•	Proper meditation and breathing techniques. There are many types of meditation. The practices of many exoteric schools of yoga, internal martial arts and spirituality still need to be scientifically organized and esoterically integrated to warrant a gradual and safe awakening of the *kundalini*. There are many steps and ingredients missing. Either the teachings are too slow or too cautious or techniques are done improperly and sometimes blindly. The Divine power and energy of the Soul have to be utilized for externalized service, not for the incarnated personality's good intentions without action. Also, too much meditation without service and external applications makes people mystical and too centered on inner practices rather than focusing on their externalized usefulness for social or world development.
•	Guidance of a spiritual mentor who understands the *kundalini* technically and scientifically is the foremost requirement. If people need fitness coaches for very simple tangible activities like body-building, how much more so for inner power development, which is very abstract. The spiritual teachers have already walked this path. Therefore, they are the best guides. Normally, spiritual teachers have vast experience in handling not only individuals, but also large groups. They also have assistant coaches who monitor and nurture new students. For beginners, spiritual teachers usually initiate important training programs for the most delicate part of awakening of the *kundalini*. In the near future, the esoteric science schools will be more externalized as institutes for this type of inner development.

We will include a few more simplified methods for awakening the spiritual inner powers and for gradual awakening of the *kundalini*. The more advanced techniques will be included in the next two volumes of *Inner Powers*.

Methods to develop spiritual powers through the safe awakening of the *kundalini*

General Method to Safely Awaken the *Kundalini*
Special physical exercise and inner breathing method
Simplified *mantric yoga* method

There are numerous techniques to awaken the *kundalini*: some safe, others unsafe and many very dangerous. A practitioner is fortunate if he or she has a found a spiritual mentor or teacher to guide his or her practice. In most cases, people copy techniques here and there, assembling them without knowing the possible side effects of combining different methods. In medicine, there are certain contraindications for mixing some medications. The same is true with spiritual practices. Some rituals and techniques are not supposed to be combined or are best performed in specific sequences. Many spiritual practitioners practice blindly with no science to it, causing their physical and psychological health to falter or deteriorate. The energy of the *kundalini* is a force that can burn energetically if mishandled. This is the reason why most teachers have not revealed this subject thoroughly for public scrutiny because of its danger to people who might experiment with it without proper preparation.

The subject of *kundalini* has long been a mystery. In this book, the techniques include basic principles to understand the nature and uses of the *kundalini* power, a few important techniques to awaken it and precautions to abide by for its safe practice and applications.

First method: Physical exercises & breathing techniques

Practice #1

(3 times weekly)

1. MDP Body Therapy and Inner Power Development Program (25 minutes) (pages 81)
2. MDP Synchronized Breathing Technique (3 sets, 5 minutes) (pages 54)
3. Nostril-controlled breathing for both nostrils (5-10 minutes) (pages 326)
4. Navel breathing and storing the energy, using navel-hand *mudra* (5 minutes). Do slow inhalations and exhalations while focusing on the navel chakras, with your hands positioned in the navel-hand *mudra*. This step develops your internal power and stamina rapidly. It will also help gradually develop your *kundalini* to a higher degree of awakening and circulate the alchemized energy in the bodies

Navel-hand *mudra*

Connect the index and middle fingers and thumbs of each hand then join them with the left ring and small fingers on top of the right ring and small fingers. Put your hands in front of your navel.

Hand *Mudras* for Meditation and Circulation of Energy (Hands Positioned at the Navel Area) Figure 35

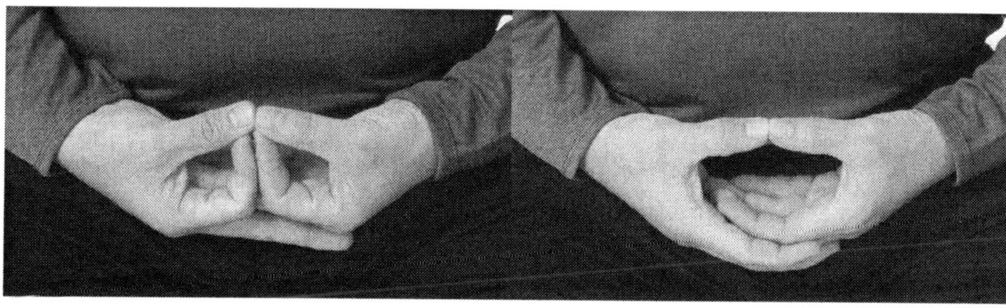

Triple Finger *Mudra*
(Thumb, Index and Middle fingers together)
Special *Mudra* for Balanced Activation of Upper and Lower Chakras and for Circulation of Energy

Dyani Mudra
(Left Hand over Right Hand)
***Mudra* used for Meditation and Circulaion of Energy**

Second method to awaken the *kundalini*: Simplified *mantric yoga*

A *mantra* is a syllable or sets of syllables repeatedly recited either mentally or verbally to produce a pre-arranged type of vibration and energy, which embodies certain types of inner powers or virtues. Most *mantrams* also heighten the frequencies of the different bodies and energy centers. *Mantric yoga* is a science of using these sacred *mantrams* to:

- Purify and revitalize the energy centers and bodies of the incarnated personality
- Develop specific virtues and inner faculties

- Fuse and integrate the incarnated personality to the Soul for ordinary practitioners or to the Divine Self for very advanced practitioners
- Awaken the *kundalini* and alchemize it with the Soul's energy
- Facilitate the circulation of the alchemized *kundalini* energy throughout the bodies
- Transform and heal oneself and others
- Call forth for the blessings of higher spiritual beings or the deliberate manipulation of natural forces

Practice #2: Simplified *mantric yoga* method to purify the energy bodies and for safe awakening of the *kundalini*

- Recitation of the sacred words *OM, Aum* or *Amen* to increase your rate of vibration and align your consciousness to the Soul. These words are also used by the major religions and spiritual groups, who consider them sacred. When recited repeatedly and properly, these words increase the rate of vibration of the energy bodies and help align one's consciousness to the Soul or Spirit.

(10 minutes, 3 times weekly)

1. Do 3 sets of the MDP Synchronized Breathing Technique.
2. Verbally or mentally sound OM, Aum or Amen with the following rhythm, selecting only one of them:

a. Ooooooooooo (80%) (sounded as “oh”) and Mmmm (20%)

b. or Aaaaaa (40%), Uuuuuu (sounded as “oo”) (40%) and Mmmm (20%)

c. or Aaaaaaaaaaa (80%) Mennn (20%)

Hold your breath for a few seconds between inhalations and exhalations.

3. For the first 2 minutes of reciting the *mantra*, simultaneously focus on your crown, ajna and heart chakras whenever you inhale. Do the inhalations slowly.

4. For the last 3 minutes of reciting the *mantra*, be aware that you are the Soul. For more advanced practioners, be aware of yourself as the Divine Self.

5. Sound the *mantra* while focusing on the navel chakras. Always focus on the navel chakras – etheric, emotional and mental levels – as you recite the *mantra,* especially during inhalations.

Verbally or mentally sound the universal sacred *mantrams* Om, Aum, or Amen for 3 minutes with the following rhythm, selecting <u>only one of them</u>:

a. Ooooo (20%) (sounded as “oh”) & Mmmmmmmmmm (80%)

b. Aaaaaa (20%), Uuuuu (sounded as “oo”) (20%) & Mmmmmmm (60%)

c. Aaaaa (20%), Mennnnnnnn (80%)

Pause for a few seconds between inhalations and exhalations.

6. Use the navel-hand *mudra* described on page 332 while doing navel breathing and the *mantric* yoga techniques.

This method should not be practiced more than 15 minutes, 3 times a week because it can overstimulate the sexual energy and may over-awaken the *kundalini* energy, which can result in physical difficulties or psycho-spiritual crises. Do not practice this method if you suffer from cancer or disorders of the intestines because it can aggravate the condition. If there is a negative reaction to these practices, stop immediately and perform at least 30 minutes of exercise, especially the MDP Body Therapy and Inner Power Development Program™. This exercise releases energy blockages, aligns the energy centers and facilitates circulation of the *kundalini* energy.

The navel chakra is the synthesizing center of the lower chakras, the sex and base of the spine. It also converts the mixture of the energies from the lower part of the bodies, especially those affecting the seat of the *kundalini* in the sex, perineum, and base-of-the-spine area. There are many minor navel chakras that can assimilate and store the alchemized energy in the navel area. In most people, there are only one or two inactive minor navels below the navel. As the person develops, especially with esoteric techniques, many minor navel chakras can be developed around the navel center. Advanced Sages have constructed many levels of minor navels, which serve as storage equipment for tremendous amounts of internal energy. These deposited energies are one of the real sources of fuel for the inner powers to be sustained. They are one of the most important esoteric tools in the path of enlightenment because many advanced faculties depend on the alchemized energy.

Precautions for the meditation and breathing techniques

In all the meditation and breathing practices and methods, the following are recommended precautions:

- Do not meditate or perform breathing techniques for at least 3 hours after a heavy meal.
- Do not take a bath or shower until 3 hours after meditation or breathing techniques.
- Do not perform the techniques more than the recommended frequency.
- If you are in doubt about your fitness or health conditions to practice the techniques, consult a physician or qualified health experts.
- It is not recommended that pregnant women, smokers, alcoholics and recreational drug users practice the meditations and breathing techniques that awaken the *kundalini*.
- Practitioners who eat red meat, especially pork, cannot necessarily safely practice awakening the *kundalini*: The spiritual teachers recommend a vegetarian diet for

advanced spiritual practitioners to achieve maximum results with minimal side effects.

The spiritual powers are safe to practice and apply as long as:

- The inner powers are used without harming others and are applied with good intentions
- One knows how to use them properly, safely and moderately
- One utilizes inner powers for personal progressive growth and service for a greater world need
- The attainment of inner powers is not the end but a tool for growing towards greater light and life
- The practitioner continues refining the personality or character to transform the character flaws that might hinder future progress and inner development or that might trigger the misuse of their acquired inner powers. Many have failed in this aspect, even more advanced human beings. Therefore, the next emphasis of this book is about culturing and refining the personality

Major sources of spiritual inner powers

Fourth source: Power of virtues through culturing of the personality

The inner powers can only be developed safely and used for greater good if the character or personality of the practitioner is cultured with virtues and universal values. We are not only talking about the harmful effects of a person's attitude towards others, but also about the danger of hurting one's growth. As mentioned earlier, when the *kundalini* power is activated and highly stimulated, it becomes a fertilizer amplifying both positive and negative tendencies of the practitioner. The extent of these tendencies depends on the ability of the person to weed out negative obstacles and behaviors when they emerge. If these vices are amplified on a large scale, it is easier to recognize them in order for the flaws to be purified and worked out. Most people normally ignore imperfections in their character when they seem minor. The Soul dispenses a certain dose of karma to be worked out yearly, and if the person delays the rectification of the negative attitudes, they may result in increased difficulties and obstacles in life. With the *kundalini* active and stimulated, these negative attitudes or flaws can become more severe.

There are several virtues and universal values to be mastered in order to be able to handle and use inner powers safely. We are using virtues and universal values almost as synonymous terms. A reciprocal relationship also exists between virtues and faculties. The process of applying a virtue is a faculty. The faculty is an instrument to express virtues and universal values. Each virtue developed corresponds to inner powers acquired and levels of spiritual development attained. The more mastered the virtues, the greater the power. Let's study the subject of virtues and put them into practical application. The following are virtues and corresponding benefits when they are mastered:

First Virtue: Benevolence

The power to achieve happiness, contentment and right human relations and the key to accessing greater inner power

This virtue is expressed as being loving, kind and compassionate in a selfless way. Benevolence is a powerful tool to maintain right human relationships and is the key to being loved and liked by many, which is one of the most powerful qualities in great inspirational leaders and motivators. This is a source of power for charismatic leaders and celebrities. At first, a person utilizes the power of the heart energy center, the quality of personal love. Then after the person has developed spiritually and the crown chakra is active, the true quality of detached or selfless love starts to be expressed as benevolence. This virtue also has many ramifications like altruism, which will be discussed separately.

How can benevolence be developed fast or enhanced further?

Steps to Develop & Enhance the Virtue of Benevolence
1st Step: Start your day by expressing goodwill. Create and project good intentions and wishes to as many people as possible daily. Goodwill is the first rung in the ladder of success and the safety valve in the acquisition of power.
2nd Step: Avoid focusing on the negativity or mistakes of others and oneself.
3rd Step: Benevolence is expressed through pleasant words and communications.
4th Step: Practice the principle of forgiveness.
5th Step: Express benevolence through active service and applied action
6th Step: Practice the spirit of fairness and a sense of justice

1. **First step in developing and enhancing the virtue of benevolence:**

   Goodwill is the first rung in the ladder of success and the safety valve in the acquisition of power.

   Goodwill expressed as good intentions is the seed of good actions, which when practiced regularly convert into good habits. Many good habits comprise a cultured personality. Therefore, positive thoughts, pleasant wishes and right motives are a good way of building benevolence thoroughly. Benevolence is not just an active way of expressing love or kindness, but is also abstinence from harmful thoughts, words and actions. Start and end your day with intentions of goodwill. Humanity is growing to implement this virtue en masse.

Technique to build goodwill

(5 minutes daily)

1. Spend 5 minutes right after you wake up to greet your day with gratitude for life and the Source of your life. Use the feeling of the heart to express this for 1 minute.
2. Also silently express your good wishes to your family and loved ones, friends, co-workers and people whom you will be meeting throughout the day. Do this for 2 minutes.
3. Visualize or have the intention of a good and successful day and see your whole day's programs and goals achieved. If you have problems, visualize them being resolved properly. Focus more on the positive outcomes of your day (2 minutes).

2. Second step in developing and enhancing the virtue of benevolence: Avoid focusing on the negativity or mistakes of others and oneself.

If you want to avoid losing much time and energy, it is better to avoid contemplating too much not only on the imperfections of others, but also on your own weaknesses and mistakes. When there is a problem, you should focus more objectively on the solutions and lessons to be learned. Excessive analysis and evaluation on your negative tendencies can lead to guilt, anxiety, resentment or depression. And over-brooding on other people's faults can result in destructive criticism, anger or resentments. It is important to remember that no one has perfect character and everyone is capable of committing mistakes, even you! Being mentally objective is important if you want to master this virtue of benevolence.

> There is nothing wrong with a green fruit. It is just waiting to ripen. Some people are like a green fruit. They just need time to evolve, and evolution requires many processes, which implies more time.

3. **Third step in developing and enhancing the virtue of benevolence: Benevolence is expressed through pleasant words and communications.**

We are not born to please or agree with everyone. But there are ways in which verbal communication and the way of expressing things can be improved. I still believe that people have to say what they really mean and mean what they say, but in the most appropriate compassionate way. In the business world, some businesspeople compromise principles and values for financial profit and because of the doctrine that the customer is always right and needs to be pleased. Usually pleasing the clients and giving the best service are important procedures, but not to the extent that we compromise truth and values. For normal people, this sidetracking of virtues seems acceptable, but as a person evolves, righteousness and virtues cannot be compromised. Practice making a habit of verbally nurturing people while expressing the rightness of things sincerely and selflessly without having to always expect gains or rewards. This is true benevolence. Anyway, goodness begets goodness and without having to claim rewards, the law of nature fulfills itself to award what is due to you sooner or later.

> Nurture others with positive, truthful words. It doesn't cost anything to do so. Mean what you say, and say what you mean. Do it everyday. This is one of the most obvious qualities of the most beloved leaders. If you cannot be generous financially, at least be generous with your words.

4. Fourth step in developing and enhancing the virtue of benevolence: Practice the principle of forgiveness.

> The old adage:
> "To err is human and to forgive, divine"
> (Alexander Pope, English poet)
> holds true.

Most of humanity needs to learn how to forgive and forget the negative past more, especially traumas and bad experiences. When people recall the negative experiences and people who have hurt them, they go into a cycle of resentment and anger, which not only drains their energy but also creates a negative emotional pattern that slows down their development.

How do we accomplish this release from the negative past?

Try this:

Releasing from the negative past technique

As if this were the last day of your life, ask forgiveness from those who have been hurt by you and grant forgiveness to people who have hurt you or are still hurting you. You might want to make a habit of the following process monthly or periodically:

1. Briefly list the people who have been hurt by you, and highlight the lesson(s) behind it.
2. List the people who have hurt you or are still hurting you, and summarize the lesson(s) behind the problem.

3. Perform 3 sets of the MDP Synchronized Breathing Technique.

4. Close your eyes and imagine yourself in front of an ocean. Inhale and exhale slowly and relax. As you exhale slowly, externalize all negative feelings and thoughts about any relationship issues as if the ocean is absorbing these discomforts. Do this for about 3 minutes or until you feel better and calmer.

5. Visualize the people who have been hurt by you, one at a time, and mentally ask for forgiveness sincerely. Express your apology and imagine yourself forgiven. Make a firm resolution not to do the mistake again and imagine that the relationship is healed, reconciled or freed from disturbances.

6. Visualize the people who have hurt you or are still hurting you, and express your forgiveness to them, one person at a time, as if this were your last chance to do so. Mentally give your advice or comments objectively, and imagine them doing the right activities or learning the proper attitudes.

7. Mentally say the following affirmation at least 3 times:

   I am a being of love and kindness. As I ask for compassion, I show compassion. As I ask for forgiveness from people whom I've hurt, I give forgiveness to anybody who has hurt me. Let all be completely forgiven. Let all be free. So it is!

8. As much as possible, do all of the above processes a few times a week to make the new positive emotional and mental programs permanent.

9. Teach this technique to others if you think they would benefit.

There are two aspects of forgiveness: external and internal. Internal forgiveness is what we just finished as a process in the above technique, but the external aspect is a physical process. The violator should be informed or educated of the right attitude. If it is a violation of the law like a murder, then there should be justice and action to discipline the violator so the same mistake is not repeated. Also, physical, emotional and/or mental repair of the relationships – or a process of reconciliation – is almost always required. But first, it has to start with the good intention to forgive. When both external and internal forgiveness are achieved, then forgetting is also achieved spontaneously. When the problem is understood, detachment follows, and forgetting the negative experience and traumas of the past is experienced naturally.

5. **Fifth step in developing and enhancing the virtue of benevolence:**

   Express benevolence through active service and applied action.

Benevolence starts with positive motives and kind intentions, expressed as nurturing words, but it should be practiced through applied action. It can be actualized as:

- Kind regards to others expressed as understanding, consideration and compassion
- Helpfulness and selflessness accompanied with being mentally, emotionally and physically supportive to people and to all kingdoms of nature

- Establishing projects and services that help as many beneficiaries as possible
- Selflessly motivating and inspiring other people and groups to become the best they can be

6. Sixth step in developing and enhancing the virtue of benevolence: Practice the spirit of fairness and sense of justice.

This is the key to inner satisfaction and inner peace. Being loving and kind is not enough to express benevolence. It has to be accompanied by fairness and the sense of justice to express the wisdom of the virtue fully. How can one test fairness and justice? There are man-made rules called social laws and codes of ethics that are temporal. There are also higher spiritual laws and universal truths and principles that are easy to grasp and implement by more advanced humans. At least for normal human beings, social law and codes of ethics are their initial guidelines for fairness, but as a person evolves, the spiritual principles and universal truths should be their guiding laws. The major spiritual law is the Law of Karma or Cause and Effect: What you sow, you shall reap. It simply means that people have to be responsible and accountable for their thoughts, words and actions. This law can be used to create your destiny in a positive way. You can predict your future or destiny of groups, organizations or nations by studying what they are doing, observing and violating according to the Law of Cause and Effect. All of their past and present track records of good or bad ways will be harvested in the future. This is what we call predicting destiny with concrete reasoning.

It is more practical to study the Golden Rule, one of the applications of the Law of Cause and Effect. This is a law of visible and invisible justice on thoughts created, words spoken and actions taken. The Law of Cause and Effect is expressed by two relevant subsidiary rules of the Golden Rule from ancient Chinese teachings and most religions. The Golden Rule, and its subsidiary rules, construct virtues, especially those of benevolence, altruism and justice. You can apply these rules through a process of role reversal to express justice and fairness.

Aspects of the Golden Rule
• Passive: Do not do unto others what you do not want others to do unto you, as taught by Confucius. This will help guide people on what not to violate against others. Use this rule to validate and test the fairness of your decisions. This is a principle of abstinence or subtraction of negative tendencies.
• Active: Do unto others what you want others to do unto you, as taught by Confucius. This is a tool to be fair in implementing actions or solutions to problems. It is a reminder to take action to do good or express the right solutions to solve problems. It is a reminder to take active roles, to implement positive steps and to express virtues rather than just a mere abstinence from negativity.

The two aspects of the Golden Rule and the Law of Karma, when applied, can solve even the biggest world problem or crises in humanity from the past, in the present or forthcoming in the future. Some problems might take longer to solve when applying these rules and principles, but the solutions are more permanent and result in a fair or win-win process for all parties concerned.

Recommended practices to develop the virtue of benevolence

1. Daily, upon waking up, treat the day as if it were your last.
2. Make good intentions and wishes for as many people as possible, starting with your family and loved ones, co-workers, employer, subordinates and clients. Visualize the goodness developed in you and in everyone. This step activates your heart and crown chakras naturally. Do this for 2-3 minutes.
3. Be fair to everyone the whole day. Role reversal or putting yourself in the shoes of others when making decisions and solving problems or negotiating is a good practice and habit to maintain fairness.
4. Be detached to the good things you've accomplished before they build up your ego. Do your duties as best as possible, spontaneously, without expecting fame, rewards or applause. Whatever you deserve will be rewarded to you in tangible or intangible ways. The Law of Karma or universal Law of Cause and Effect fulfills itself whether you ask for it or not.
5. Periodically evaluate yourself, especially on your birthday or at the end of the year. Do the following steps: List things that violate or harm people like your thoughts, words, habits or attitudes. Make a firm resolution to correct them and make sure you check yourself constantly not to violate the same offenses.

6. Visualize or create the intention that you are already expressing the right attitude. This step requires repeated visualization to program your consciousness to instinctively express the proper behavior in the future.
7. Open your heart and let harmlessness your natural state. The true esoteric requirement of the mastery of inner powers is harmlessness.

> Greater powers are permanently achieved through the virtues of benevolence and harmlessness. When you have attained a level where even if you are pushed to the edge, you will not misuse power, then true inner powers are at your disposal.

When the heart and crown chakras are active and esoterically open, a person becomes more inclusively loving and harmless. Therefore, activating these chakras is an important technique for the mastery of power.

8. **Do the MDP Synchronized Breathing Technique** with extended inhalations and exhalations focusing on the heart and crown chakras at the etheric, emotional and mental levels.

(5-7 minutes 3 times weekly)

Second virtue: Altruism

The key and power behind prosperity and sustained abundance

The key and power behind prosperity and sustained abundance is altruism. To be altruistic is not just being generous at the financial or material level, but also at the spiritual, mental, emotional and energy levels and in a more unconditional, unselfish way. At first, people are generous with some hidden selfish intentions in the back of their minds. This is excusable and understandable in ordinary people because they are still "spiritual children" whose solar plexus chakra is bigger than their heart center. When the heart energy center is active and functional, as in more advanced mystics or loving people, the emotional instinct of being altruistic is already a habit. The higher form of altruism is developed when the crown chakra is active and functional wherein the person gives unconditionally at the material, emotional, mental and spiritual levels. These people express generosity as a sense of duty or a natural responsibility to humanity or the world, like many rich people supporting philanthropic work.

Altruism is expressed naturally by advanced Souls, Saints and Sages because of a long-standing habit of generosity. Abundance in life is not only measured in financial and tangible material assets but also in many intangible ways like good health, happiness, peace of mind, smooth sailing projects and life's fulfillment. There is such a thing as invisible prosperity gained through the habit of altruism. You may call this good luck or good karma of entitlement. The good karma generated through generosity can be converted and harvested through different channels. It is a matter of where this good luck or karma of abundance will go or defining how it will be received.

> Good luck and the ability to be entitled to good things
> go to the people who work for them.

There are different levels of expressing generosity and altruism, starting from the merely physical and material levels. When people are financially generous, they tend to become wealthy. Financial and material prosperity are usually the result of not only hard work, but also generosity and abstention from stealing and unfairness. Hard work is not enough to make a person or organization successful and prosperous, or to sustain abundance for a long time. There is also a need for intelligent hard work. Intelligent hard work equals hard work in a well-organized manner, plus good karma of entitlement.

The speed of achieving goals and removal of obstacles are directly proportional to the amount of karmic entitlement available to the individual or the organization.

The inherited prosperity of many wealthy families and lineages were earned by their ancestors who practiced some of these principles of entitlement. But when their descendents cease to be generous, the flow of prosperity is terminated. Much inherited wealth in families and corporations fades away because of this. Therefore, it is wise to pass to the next generation or successions of leadership the virtue of altruism as the key to sustained abundance. Islam and Judaism tend to implement the law of tithing or appropriating a percentage of income for donations to the needy. Most of their members uphold and apply this principle of giving religiously, and this explains why many of them are super-rich and continue to enjoy abundance.

Taoism and Christianity also encourage altruism in their membership, but not with strict discipline as the former two religions. Currently many business organizations and wealthy individuals are becoming more generous and appropriate large sums of financial support to the needs of the world. They practice philanthropy not just for the benefit of tax exemptions on their donations, but they feel the sense of responsibility for human development.

> The individual, corporations, organizations and nations who will be wealthy or who will sustain their abundance will be those who practice altruism. This is the key to sustained prosperity, and this is the principle of good luck and entitlement.

Many spiritual teachers recommend donating an average of 10% of your net earnings. Although charity begins at home and in the neighborhood, it is recommended to extend generosity globally. I think the altruism of the United States, Switzerland, Germany, Canada, Saudi Arabia and other wealthy countries is the foremost key to their sustained prosperity aside from their technological and business advancements, hard work and persistence. These countries also have many generous people and groups. Some of the companies who pay their workforce lucratively and have made their leaders and managers millionaires are the most successful wealthy organizations today.

How can we practice the virtue of altruism better?

Those who have financial challenges need to practice this virtue more than those who are already prosperous. If you do not have this habit of generosity or if you just give once in a while, here is a recommendation:

Techniques for practicing the habit of altruism

1. **Form the habit of giving financially to a good cause.**

   Determine the cause or project that stimulates your compassion and give donations monthly. Start with an amount that you feel comfortable with according to your budget. When the habit of generosity is formed after a few months, increase the amount of tithing to 5-10% of your net income after tax. The more you need to increase your income, the more you need to give. It takes some time for most people to form good habits. It is a good idea to put dates for your monthly donations and amounts in your planner or calendar as a fixed reminder.

2. **Observe the difference in your life, materially and spiritually, when you give a lot versus when you give little or do not give at all.**

   This can be both experimental and experiential for you. Experiment over a few months to prove the principle of generosity. Practice the virtue and be patient to wait for the results. It is like the law of physics or mechanics or law of gravity. Whether one understands them, the natural laws fulfill themselves. This is true of the virtue of altruism fulfilling its cause and effect under the Law of Karma. Be scientific but patient. Do not expect the return on your karmic investment immediately. When you plant corn, you don't harvest the next week, but once you harvest after a few months, the reward is given many-fold. The law of giving generously does not follow the Law of Subtraction, but the Law of Multiplication of investment returns on the donation.

3. **Teach the virtue to others.**

   After a few months of practicing generosity and having proven it successful, teach it to others. Start with your family, including your children. Then do it with your organization. Altruism is a good value to sustain your family's fortune or a company's continuous profitability.

Techniques for already generous people

People who have been practicing the virtue of altruism as a way of life can still enhance its expression further as follows:

First technique to enhance the virtue of altruism

1. Give what you need most. Altruism is not just on the material or financial levels. You can also give and receive emotional, mental and spiritual benefits. Many people are rich financially and materially, but are not in good shape in relationships and spiritually. When people need more respect and love, they should give more respect and love.

> The thing that you need most is
> the thing that you need to give most.

Here's a simple strategy to give what you need most:

List aspects and attributes that you need most:	Individuals or projects of your life that need these aspects or attributes:
1. ____________________	1. ____________________
2. ____________________	2. ____________________
3. ____________________	3. ____________________

Make a schedule and strategy to practice altruism based on the above list. When? Immediately and constantly. It is better to start small than never, therefore, start giving or sharing whatever you can as soon as possible.

2. **To whom do you give?** Invest your donations in the right cause with the right purpose. Giving and tithing should be done wisely and properly. Donate where there is maximum benefit.

> Plant good seeds in the most fertile ground.
> Invest donations in the same way.

Second technique to enhance the virtue of altruism: Choosing where to donate for maximum benefit

1. **Determine what individual projects or world causes can maximize your donation or tithing.**

> Maximum benefit in harvesting good karma goes to a donation that has the greatest influence on world evolution and human development for the longest period of time and for the largest number of beneficiaries, human or otherwise.

2. **Give to people, groups or organizations that have helped your life directly.** Every year, make a list of people or organizations you can include in your practice of altruism yearly. The amount of tithing is directly proportional to the help given to you and the necessity of the individual or organization.

3. **Invest in good karma through donations before your need arises.** It is easier to concretize your goals if there is entitlement to achieve them. This entitlement is lubricated by the good favors you've given to the people or

organization involved in the project, but foremost is the good karmic account you've invested in ahead of time. That is why when people receive things beyond their expectations, we say that he or she is fortunate. This is connected to the good luck or good fortune mentioned earlier as invisible prosperity. List those areas of life you need to enhance and invest your donations in a related world necessity as follows:

Personal requirement	Areas to invest donations
Family success and future of children	Projects or organizations that help families or orphans and children
Good health	Projects or research on health and enhancement of health for others and humanity
Enlightenment and mental development	Work and projects on enhancement of education and scientific research
Good relationships	Research and projects on the improvement of human behavior and quality of life and resolution of conflicts, including works of the United Nations
Spiritual growth	Projects and organizations that help with the inner development and transformation of humanity
Beauty and attractiveness	Projects that clean and beautify the environment and also in works to improve arts

Self-interest in the return on your good karmic investments is not against altruism. It is an intelligent way of using resources to be able to receive more in order for one to have more resources to give. As long as the intentions are good and the self-interest is not hurting people, groups, organizations or the world, it is not classified as selfishness. Having self-interest is not necessarily selfishness as long as the motive is right and the action to harvest rewards is harmless and not at the expense of others. It is still considered a virtue.

Other levels of altruism

Living life is not just about material success and financial reward. As you evolve, practice and integrate all the levels of altruism as follows:

- **Emotional altruism**: Nurture people with good and pleasant feelings and expressions of emotional kindness. Utilize love, compassion and understanding as a tool. Do it daily to as many people as possible

- **Mental altruism**: Sharing good ideas with others, even with competitors, is the key to mental enlightenment and the good karma of receiving better ideas. This might raise a few eyebrows among those who consider it a mistake. But if you want to receive new higher ideas, release the old ones that you have mastered and let others benefit from them. Teaching others the topic you want to master is the key to becoming an expert on the subject matter. Mentally nurturing and motivating others to become better is also an expression of this mental altruism. Even good intentions and wishes mentally created or expressed are considered mental altruism.

- **Spiritual altruism**: The key to rapid spiritual growth is to be spiritually generous to others. Supporting people to grow spiritually and facilitating their ability to comprehend their life's lessons and fulfill their lives is an expression of spiritual generosity. Giving advice on spiritual values and wisdom to others, especially to those who need timely help, is recommended. Therefore, supporting your religion or spiritual mentors and their organizations to spread teachings to transform and enlighten humanity is a great duty and service, which is usually the responsibility of more advanced human beings. Another example is to be compassionate and forgiving to people or organizations who cannot fulfill their financial obligations to you, as in the case of bad debts. Not only will we forgive and cancel the loan, but perhaps we will help people or organizations recover. This might be too much for some people, but this virtue can be applied more easily by more evolved souls.

One thing I'll never forget as an example of this virtue is a man I met in Cali, Colombia: His teenaged son was in the final championship of a bicycle race, but before the final lap, a man murdered the son. It took the father a few months to recover form this trauma, but one day he regained his spiritual sense and internally forgave the murderer. He later found out that the killer, who was put in jail, also had a teenaged son. One day the father of the murdered boy visited the murderer's son and offered him his dead son's bike. He proposed to train him for the next bicycle race. The murderer's son won the bicycle race the next year with this gentleman's help and coaching.

How big is that heart and how strong the courage to implement the true Christian spirit that if they throw stones at you, throw them bread? This is spiritual altruism at work.

Things to avoid in order to unblock the flow of prosperity
• Do not delay cash flow or withhold what is due to others.
• Do not look down on lesser people, like the poorer segment of society.
• Avoid pride. The lesson of pride is to be humbled, including in financial status.
• Avoid excessiveness and fanaticism. Rich people and big corporations go bankrupt when they violate this principle. It is natural that as you climb the ladder of fame and success, overhead expenses also increase – most of the time exponentially. But know your limits and always save, not only for the present, but also for the rainy days.

The virtue of altruism is first formed by small doses of generosity until it can automatically express as greater altruism up to the emotional, mental and spiritual levels. Like any other attitude, it takes a good reason to develop it and a lifetime to master, but it is never too late to do it, so practice the concepts immediately. Make the positive intention today and the action to do it follows.

Third virtue: Discipline and constancy

The key to attaining greatness and achieving big results

> When one has discipline and constancy,
> success is predictable!

This virtue is the mark of a champion and winner either in the Olympics, business, spirituality or in any of life's endeavors. Why are some people more successful and have achieved more in less time? Not just because of talent, because many talented people have failed repeatedly in history, nor is it only technological advancement because many scientific and technology fanatics also became failures. It is the discipline and the constancy of effort that is the single most important factor that propels people to attain greatness and achieve big results. The virtues taught by most religions and spiritual schools emphasize the aspects of love, compassion, generosity and truthfulness, yet seldom do the practical aspects of virtues and values get emphasized.

On the other hand, competitive sports, martial arts and military training have always embodied this virtue, which produces the best athletes in the Olympics, the most skillful martial artists and the most decorated soldiers. Due to the force of competition for excellence and future promises of financial reward, university students also develop this virtue, which later on is mastered by scientists and businesspeople throughout their careers. The people who have mastered the virtue of discipline and constancy become super-achievers. To practice true discipline and constancy the following are recommended:

Techniques for building the virtue of discipline & constancy

1. **Set your targets and goals clearly and know by heart your most important roles and end results**. Make sure that your goals are important and form a good part of your life's vision. It is not enough to be constant in effort if the direction and purpose of the goal to be accomplished is not positive, certain or clear. Know your goals and required results, qualitatively and quantitatively, by writing them until you are clear and definite about them. Constancy of focusing upon them in good or even in bad times is the key to discipline. Remind yourself of your goals and results in your

calendar or planner and use props as reminders in your house or work place. Put them into a slogan or motto. An organization or team can do the same.

2. **Mastery of any work or endeavor needs more time than normal and the force of discipline.** In general, the thing that you have done the best is what you put more time, energy, resources and effort into with constancy and unwavering focus.

> Remember that the most successful people are those who are able to go the extra mile in effort and spend more time doing the most important things that bring the greater value to life.

3. **Make a practical strategy and schedule in writing that specifies the exact dates and time needed to accomplish the goals, and accommodate the activity that produces the required results.** Follow the schedule and deadlines you set for yourself strictly without compromise. Writing and knowing the schedules and results are not enough to be disciplined. You need to implement strategies and monitor your results. Adjust accordingly based on your periodic evaluation.

4. **Finish what you start as much as possible**. For most achievers and successful people, this is already a habit, but under-achievers need to work more on this aspect. You can start on smaller and short-term goals and projects that inspire you and get into the habit of finishing what you start. If you have done this several

times successfully with smaller targets, then doing larger goals is less and less difficult.

5. **Focus on the results, but do not compromise the virtues in implementing the process**. Sometimes people sacrifice the processes and procedures just to achieve what they want, even to the extent of violating virtues. It is counterproductive in the end. Just like athletes who take steroids to enhance their performance in the Olympic Games, but end up losing it all when their medals are revoked. Or some businesspeople or politicians might win and achieve their goals in the short run, but experience bigger failure at a later time. Real discipline takes into account the right virtue and process to achieve the best results. Do not forget that your goals are just steps to long-term plans, and it is not the end of the world for you if you do not achieve them. After you do your best, be detached to whatever results follow and accept them. It is your attitude, after all, that matters.

6. **Work to achieve best results in the shortest time, but make sure you survive**. It is better to be a successful winner or champion who is alive than dead. Do not sacrifice your health or life just for one goal you set. You are not the goal. It is just one of the targets you have created. The goal is not you. You have more things to achieve than one goal. Therefore, avoid being excessive and causing loss of life or great discomfort to others in pursuit of goals.

One should not mistakenly interchange courage and foolish fanaticism. These might be similar qualities, but they are not the same attitude.

7. **List all attitudes and obstacles that negate your goals and results, and work them out.** Plan contingencies to solve expected problems and avoid surprises. Develop the next skills, aptitudes and attitudes needed for your next steps or goals.

8. **Discipline is best tested and developed where there is a devil's advocate or the presence of competition**. Therefore, implement goals with a sense of urgency. Create a sense of healthy competition if you need it to perform better.

9. **Do not rely on your own natural capabilities**. For important goals, get a coach or mentor who has been there, a wiser person or an expert. Select your coach or mentor based on their track record.

10. **Change or adjust your goals and targets if they are destructive or counterproductive to your life.**

So how else can you enhance the virtue of discipline and constancy?

Energy centers that enhance the virtue of discipline and constancy	
Base-of-spine	Physical stamina, strength & practical groundedness
Navel and minor navel chakras	Vitality, staying power & instincts
Solar plexus	Determination & emotional commitment
Heart chakra	Emotional stability & inner peace
Ajna chakra	Mental stamina & will-power
Crown chakra	Wisdom & spiritual will-power

Esoteric method to improve the virtue of discipline and constancy (3 times weekly)

Do the MDP Synchronized Breathing Technique (5 minutes).

1. Focus on the following chakras and their 3 levels – etheric, emotional and mental, and slowly inhale and exhale. Hold your breath for a few seconds in between inhalations and exhalations. Do 10 slow breathings for each set of chakras:

   a. Crown
   b. Ajna
   c. Heart
   d. Solar plexus (8 times only)
   e. Navel (Avoid this step if you are pregnant.)
   f. Base-of-spine (Omit this step if you have cancer, high blood pressure or are pregnant)

In these modern times when life's pace is so fast and the demands and challenges are more complex, mastering the virtue of discipline and constancy is a great advantage. This is a virtue that makes the great leaders, successful business people, extraordinary athletes, brilliant scientists, Saints and Sages excel in whatever endeavors and goals they embody. Not all of the above groups have mastered benevolence but all of them have discipline and constancy of focus and effort in order to be successful.

Fourth virtue: Objectivity and practicality

Power of discrimination to achieve balanced results – the antidote to fanaticism

We are finishing up an era during which the martyr complex is common in individuals and groups, and fanaticism through excessive idealism and blind faith are the guiding attitudes of many leaders and their followers. It is a dangerous world where excessively fanatical sects and militant groups are busy implementing their religious or nationalistic ideals. How did this come about? What is the antidote to this fanaticism and excessive idealism? There must be a flaw in certain educational processes or some religious teachings that triggers the lack of objectivity and practicality in many people. The virtue of objectivity and practicality develops the power to know and perceive the truth properly and the ability to adapt a method or action that is suitable and appropriate to the situation. It is an attitude of being realistic and grounded.

How can one develop the virtue of objectivity and practicality?

As there are many levels of truth, you need different levels of faculties to perceive truth and information. Then after concluding the facts and data properly, act upon and use them right. Perceiving truth accurately is the first step, and expressing the truth or information in a practical manner is the second step. Many brilliant people research information and conclude the facts, but do not use the data for practical application. What a waste of time and data! On the other hand, many practical people perceive the right information, but due to lack of wisdom, express the data wrong and destructively. Both accurate and objective perception are musts, and practical application is required.

Methods to master the perception and validation of truth

1. If there is insufficient data in a very urgent situation, use your physical instincts. This requires the development and instinctive powers of the navel and base-of-the-spine chakras.

2. If you have developed emotional instincts, use your gut feeling to validate the rightness of the perceived data, information or situation.

3. Use your emotional intelligence or inner conscience to sense what is right. This is a function of the heart chakras.

4. The concrete mental intelligence for analysis has to be used properly to discriminate what is right and wrong and what is ideal or practical. Verify all the above findings from steps 1, 2 and 3 using the process of concrete reason. The throat chakras are employed for this faculty.

5. Evaluate all the above findings by comparing them to existing common sense. This is the function of the throat chakras and the concrete mind. Scientific data are still the most reliable sources of common sense.

6. Since not all advanced information and events are known through common sense, evaluate the situation or data

employing the abstract mind. This requires abstract principles and philosophies to validate the information. The questions "why?" and "what for?" are used to probe deeper into the issue or situation. The ajna chakras and the abstract mind are utilized for this faculty.

7. For more advanced human beings who have partially developed the faculty of intuition or have sufficient intuitive awareness, perceive the idea or situation and sense if it is coherent with bigger and higher ideas. Evaluate if it lends a sense of integration with greater truth. The crown, forehead and ajna chakras are required to develop this faculty, with the help of the Soul's wisdom.

8. Another tool to evaluate the objectivity of data or information concerning philosophy is to test the universality of the idea or information. How inclusive and encompassing is it? Does it foster separativeness, or does it improve synthesis or coherence with a greater Truth? Does the data violate the interpretation of holy scriptures or universal teachings? Consider the knowledge and presentation of truth from the Esoteric Sciences or from the teachings of the Ageless Wisdom. This is not necessarily connected to the so-called "New Age" interpretation and information. Be wise and selective in choosing references.

9. If, at the end of all these tests and evaluations of truth, you are still doubtful, consult an expert on the subject matter. If you are still uncertain, get a second or third opinion from other experts.

10. At the end, synthesize all the gathered data and opinions and decide according to the best discriminating faculty you have developed. This might be tedious at first, but when you practice this protocol, especially by briefly writing the evaluations, you will develop the power of a scientific research attitude and real objectivity. The more you use the different levels of validating truth, the more secure and certain you are in your discrimination and perception. These steps and procedures will be elaborated on in Volume II of *Inner Powers*.

Guidelines for digesting information and expressing practicality

1. In general, it is not necessary to look for the absolute truth in an idea or information, but to evaluate what components of the data are correct, acceptable or not valid. This is where people have difficulty because they are either perfectionists or too idealistic. Sometimes you need to be patient like a miner finding the ore of gold or diamond in a mountain. The same is true with great ideas or information: We need to look patiently for the precious part of the idea or data. Know the limitation and percentage of probability and potentiality of the truth, idea or concluded information:

    a. What percent of it is possible? What are all the possibilities?

    b. What percent of it is probable? What are all the probabilities?

c. What are the limitations and potentials?

d. What elements can be wrong? What aspects are valid?

e. What are the seeds of success of the idea or chosen situation?

f. What are the seeds of failure of the idea or situation?

g. What can go wrong? What are the potential problems?

h. What are the worst things that can happen? Can they be handled safely or properly?

i. What are the positive and negative effects on others and the environment?

j. Does the perceived information or situation support or negate universal values and higher principles in life?

2. Express the perceived truth, idea or situation in many practical ways that benefit you and others. Write and answer the following:

   a. What are the potential ways to use this data, information, truth or situation? Choose the best ways to use the truths or situation.

   b. What are the benefits for others?

   c. What are the positive and negative consequences?

   d. How many are benefited and to what degree?

   e. How long will the positive or negative effects last?

f. How do I start to use and apply this data? What are the strategies?

g. What are the contingencies and probable solutions for any foreseen obstacle or problems?

h. Create a short-term or long-term plan based on the questions and information above.

i. Study and apply procedures from Section C on Mental Powers, how to materialize ideas (page 125), how to solve problems (page 140), how to concretize and organize life's goals (page 144), how to physicalize projects (page 145).

j. Synthesize all your best answers and devise a practical plan to apply the right information, data or situation. Select one or more practical ways that can benefit your life and improve the lives of others. Make sure you finish what you've started with the highest quality results and a sense of urgency. The guidelines above are specially designed for bigger and more complicated situations or for more complex fact-finding endeavors.

k. After implementing your decision, monitor results and summarize them in a report.

In general for daily simple situations, practice applying objectivity and practicality as follows:

Methods to apply discrimination & practicality

1. Briefly relax and center yourself for a minute, if stressed. Use the MDP Synchronized Breathing Technique; even 2 sets will suffice.
2. Use your physical and emotional instincts to perceive information or a situation fast.
3. Evaluate the perceived data using your common sense and concrete reasoning faculties. Check if the data or part of the information sound right.
4. Check the data or situation to see if it is good for your life and for others.
5. If it is good and beneficial, go ahead and apply it immediately. If it is destructive and harmful, forget it and drop it immediately.
6. Make sure it is in line with your life's important goals and is not only beneficial for the present, but also for the long term.

Objectivity and practicality is a necessary virtue to produce great outcomes in life, and it is the key to avoiding wrong decisions and to prevent fanaticism. Being benevolent, altruistic, actively creative and intelligently grounded to do the right actions creates balance in life.

Esoteric method to develop the virtue of objectivity and practicality

(7 minutes, 3 times weekly)

1. Do 2 sets of the MDP Synchronized Breathing Technique for 3 minutes. Do not do this if you are pregnant or suffering from heart conditions or high blood pressure.
2. Focus on your ajna, navel and base-of-spine chakras while breathing slowly about 2 minutes. For people with hypertension or who are pregnant, only focus on the ajna and navel chakras.

This technique is simple but it helps you develop the virtue of objectivity and sense of practicality if practiced regularly.

The Lord Buddha was a master of discrimination. He said:

> We must not believe in a thing said merely because it is said; nor traditions because they have been handed down from antiquity; nor rumors, or such; nor writing from sages, because sages wrote them; nor fancies that we may suspect to have been inspired by a *deva*; nor inferences drawn from haphazard assumptions we may have made; nor because of authority of our teachers or masters. But we are to believe when the writing or doctrine or saying is corroborated by our own reason and consciousness. For this I taught you not to believe merely because you have heard, but when you believe of your consciousness, then act accordingly and abundantly.
>
> \- Lord Buddha, the Enlightened One

Some mystical and religious groups and idealistic people may want to incorporate the virtue of objectivity and practicality in their daily decision-making and problem-solving. Mystical people are very good at emotionally and intuitively perceiving the truth. In fact, many of them are psychics and visionaries and recognize situations or information ahead of others. But it is the sense of practicality of using the information and the ability to react properly and in a timely manner to the events that has to be enhanced. The lower chakras, especially the navel and base-of-spine centers, have to be activated and revitalized for more practicality and groundedness. The idealist should act upon their ideas and ideals immediately and make these ideas useful in many ways. Talking and preaching about the goodness and greatness of ideals is not enough. Walking the talk is more important.

> **One good idea applied immediately is better than five memorized and stored unused in the mind.**

Businesspeople are good at practicality and scientists are good at objectivity. Adam Smith was precise in his statement:

> **Science is the great antidote to the poison of enthusiasm and superstition.**
> **- Adam Smith, Economist**

By developing the inner faculties and powers through the techniques in this book's many sections, we can integrate, master and externalize the objective mind of a scientist, the heart of a Saint and the practicality of a businessperson. The virtue of objectivity and practicality increases your performance to produce great results with certainty and converts your efficiency to effectiveness.

Fifth virtue: Will-power and vitality

Key to speed and continuity of long-lasting performance

Many people perform great miracles with the virtue of benevolence and altruism; others do well creatively with the virtue of objectivity and practicality, but without will-power and vitality, it is not easy to maintain performance at a greater speed and to continuously sustain effectiveness in life. Without the virtue, it is impossible to master the virtue of discipline and constancy, it is difficult to overcome difficulties and obstacles and people tend to preserve the present and dwell on the obsolescence of the past. It is will-power that pulls people out of the past and pushes them to the future. With will-power and vitality, benevolence and altruism can be expressed naturally and continually. When people are low on vitality and will-power, it is difficult to be effective in life. It is also not easy to be nice and pleasant when one has a low battery. Will-power and vitality are best expressed as follows:

- Vitality is the fuel for greater performance.
- Will-power is the 4-wheel drive to boost the power of performance and effectiveness in life. It is the dynamite that allows you to blast through great obstacles and difficulties.
- Will-power is the pioneering force that opens and constructs your road to success where others have not gone before.

> The levels of your vitality and the force of your will-power are directly proportional to your speed achieving results and the levels and quality of your success.

How many people get so frustrated because they fail in the last stage of their projects due to the lack of power at the end? Many usually start strong and enthusiastic, but their will-power runs short and vitality goes down before finishing projects. What makes great champions and winners is not just the starting phase of the game, but the end game. This is true in chess and in sports. It is also true in business performance and personal relationships. The winning edge and the ability to achieve great goals require extraordinary vitality and sustained will-power.

Let's look at levels of will-power and the chakras and energy body(ies) employed:

Expressed level of virtue	Bodies employed	Chakras required
Stamina & staying power	Physical & Etheric	Navel & Base-of-spine
Determination & commitment	Emotional	Solar plexus
Persistence & mental stamina	Mental	Ajna
Spiritual will & unwavering persistence	Mental & Soul	Crown
Integrated will-power	Integrated bodies & Soul	All of the above
Divine or supreme will-power	Soul will or Divine will	*Atmic & Monadic*

Your level of spiritual development and the degree of the Soul's or Spirit's fecundation over your personality determine the amount of will-power and vitality available to you. The more advanced the person, the greater is the virtue of will-power and vitality. This is the reason why Saints and Sages can handle bigger work and sustain projects that require global scope and thousands of years of effects and influence. Their names and impact of their ideas and teachings still sound fresh and applicable after thousand of years.

How can you increase or enhance your will-power and vitality?

We taught many techniques earlier in Section A for vitality management and some methods to improve your will-power. Let's summarize some of them:

Techniques to improve will-power: Choose any method that suits you.

A. Exercise and inner breathing method

(3 times weekly)

1. MDP Body Therapy and Inner Power Development Program: You can refer to page 54 in Section A or follow the video.
2. Breathing techniques to boost vitality: any of the 2 techniques (3 times weekly):
   - Slow abdominal breathing (first type) (page 77)
   - Combination breathing (page 81)

B. Squats method

1. Alternate nostril breathing technique (3 times weekly) (pages 326)
2. Squats exercise (20-30 squats, 3 times weekly) (pages 56)

C. **Body squeeze method** (3 sets, 3 times weekly)

1. MDP Synchronized Breathing Technique (3 sets)
2. Pushing and pulling exercise (3 sets)

This method involves inhaling fully and holding your breath while squeezing all parts of your body to increase vitality and will-power. If you want to lose weight in particular parts of your body, you can focus on squeezing these parts more than others. While holding your breath, there is more oxygen absorption in the lungs and blood. More spiritual energy also comes in as a result of inhaling and holding your breath. When you exhale and hold your breath, more *kundalini* flows throughout the whole body.

Procedure (follows step 1)

a. Stand with your feet apart at shoulder width.

b. Shake your body continuously in rhythm, loosen up and decongest blocked areas and parts (3 minutes).

c. Maintain your foot position, but bend your knees into a stance as though you were riding a horse.

d. Place your palms on your waist, palms open upward and fingers pointing horizontally towards the front.

e. Inhale deeply and fully and hold your breath for a few seconds.

f. Push your palms forward until they are fully extended directly out in front of your chest. As you are pushing your arms tensely, squeeze your entire body and exert force as if you were pushing a car. After full extension of your arms and breathing out all your air slowly, relax your whole body while your arms are in an extended position. Breathe slowly.

g. With your arms fully extended in front of you, breathe deeply and fully and hold your breath for a few seconds.

h. Squeeze your entire body and pull your hands as if you were pulling a car towards you. Finish pulling when your palms are back in their original position.

i. Exhale slowly and relax your whole body with your hands on your waist. Breathe slowly.

j. Repeat steps e-i 2 more times.

The first time, do only 2 sets of pushing and pulling. It is a very intense exercise and boosts your vitality and will-power fast. The following people should not practice this technique:

- High blood pressure patients
- Pregnant women
- People with heart conditions or chest pain
- Migraine or headache sufferers
- People with hernia
- People with arterial problems
- Unhealthy elderly people or anyone over 50
- People with high stress
- Smokers, alcoholics and drug addicts

Consult your physician if you are in doubt about your condition before you embark on this method.

D. Chakral breathing technique

(10 min, 2 times weekly)

1. Do the MDP Synchronized Breathing Technique (3 sets for 3 minutes)

2. Do chakral breathing as follows:

   a. Focus on your ajna, navel and base-of-spine chakras while breathing slowly for 2 minutes. Use only the ajna and navel chakras if you have high blood pressure, pregnant or cancer or are pregnant.

   b. Focus on the heart, ajna and crown chakras while breathing slowly for 2 minutes. Use only ajna and crown chakras if you have heart conditions.

   c. Absorb and store energy in your navel area. Focus on your navel and minor navel chakras around the navel (intention is good enough) while breathing slowly for 3 minutes. This is an important step to absorb internal power that sustains vitality and will-power.

Culturing will-power takes more time than increasing vitality. Almost anyone can boost vitality quickly, but not always will-power. Will-power depends on the development of certain chakras like the ajna, crown, navel, solar plexus and base-of-spine. But it depends even more on the substance available in the different bodies. The degree of development of the *kundalini* determines the person's will-power. The virtue of will-power and vitality also depends on how much of the Soul's will-power can come down to be utilized by the personality's chakras and energy bodies. When people neglect to develop the will-power, the Soul usually creates strife and more difficult situations for the person to develop the will.

People born in poorer countries where manual labor and more struggle is an everyday part of people's lives develop more will-power. Children born to poorer families usually have more avenues to develop their will-power than those from richer families who live more comfortably. Developed countries where life is more comfortable and people live in abundance tend to be weaker in will-power because people do not get exposed to comparatively minimal daily difficulties and struggles. I would not be surprised if nations with maximum comfort and minimal stretch of will-power are challenged with some future difficulties in order to balance their development.

The virtue of will-power and vitality is necessary to improve performance and quality of life. It accelerates the speed by which goals are achieved. It helps a person move towards the future with greater productivity and effectiveness. It is a necessary inner power that leaders, parents and groups need in order to fulfill their roles and goals. Will-power and vitality are the marks of a true long-lasting leader.

Sixth virtue: Group consciousness

The big key to sustaining power and achieving bigger goals in less time

One of the most important requirements for thriving successfully in this highly competitive modern society is the ability to work in an effective team or with a powerful group. This is true in a family, business or spiritual service. With group work, bigger goals can be achieved in less time than if projects are done individually. There are also plans and missions that require many different talents and skills that aren't often all possessed by one person. Behind the great success of any organization or leader is a group or team that made it happen. A great leader can achieve the impossible with a powerful team. It is a must in the working and development of higher inner powers.

What are the requirements of this virtue?

1. First requirement for the virtue of group consciousness:

The spirit of interdependence and coexistence

Through the spirit of interdependence and coexistence one realizes that one cannot exist without others, that the true nature of life and creation is characterized by co-existence. When this is understood, then the true understanding of group work and group consciousness starts to flourish.

> The real freedom is not in being independent from your needs of others, but in knowing that your life exists because of others.

Whether you like this idea or not, nature is like this and it will always prove itself in this fashion. With objectivity and humility, this principle is easy to understand.

2. Second requirement for the virtue of group consciousness:

The spirit of cooperation

Before a person wants to cooperate, they first know the goal and purpose of the undertaking; that the project and work are worth the investment of time, talent and effort; and that it is aligned to their purpose or goals in life. Therefore, a person has to determine the ultimate objective of the project and that it does not violate values or established codes of ethics. It is best that the individual members know the expected results of the project from the outset.

3. **Third requirement for the virtue of group consciousness:**

Proper skills and aptitude

The talents and qualifications of the cooperating person should fit the requirements and expectations of the group or leader. This will ensure that the individual member does not become useless or a liability to the group. Developing the group's required aptitude and skills to accomplish the best quality result is the responsibility of the leaders and so with every member.

4. **Fourth requirement for the virtue of group consciousness:**

Right attitude

Whereas aptitude development requires the member to have skills and talents, the attitude requirement is more of the right psychological and spiritual behavior towards the goal and within the group process. Although it is necessary for the members to cooperate whole-heartedly and make the execution of the plans and strategies easier, the right attitude does not imply that cooperation is always "yes sir", "yes, ma'am" behavior. The right attitude involves more of a freedom to express creative ideas and clear communications properly and appropriately as long as these proposals are reasonable and do not delay the projects and negate results. Once the group decision is made, everyone has to cooperate, respect the final decisions and fully implement the plans without setbacks and antagonism over their personal preferences. Fanaticism and excessiveness should be avoided, and objectivity and practically should be employed.

5. **Fifth requirement for the virtue of group consciousness:**

Proper communication

This is one of the most important aspects of group consciousness in order for teamwork to succeed. All the visions and goals have to be known and communicated clearly to the members of the group, guided with a set of clear rules and protocols. The roles and job description of the leaders and members have to be properly communicated, preferably written in a coherent and integrated manner. Openness and integrity in the communication process is a must and have to be implemented wisely. Results and performance should be monitored with reports. The group has to be organized with schedules. Small or big groups always require targets and schedules, including deadlines. It is expected that members are accountable and are trained with initiative and a mature sense of responsibility. Members should know the consequences of achieving and not achieving targets and deadlines.

6. **Sixth requirement for the virtue of group consciousness:**

Inspiration

Whether dealing with a short-term or long-term plan, groups perform better when they are inspired, happy and enjoy what they are doing. It is not the leader's responsibility to make the group happy or to inspire every member. Each group member should do their best to stimulate inspiration and the sense of joy in the group. Leaders can maintain inspiration in the group by helping everyone understand that each member's work or contribution is important. That the members make a big difference and that the group's result is justified by each member's effort. The group members and leaders should focus more on the end result and solutions rather than problems, if they arise. It is expected that plans may encounter obstacles, opposition or delays, but the important thing is to not dwell on the negative issues, but rather to pass through them courageously and objectively. A group is tested not in good times but in bad times. The group has to have a plan and contingency for possible problems to avoid surprises. In bad times inspiration and will-power are needed most.

7. **Seventh requirement for the virtue of group consciousness: Discipline**

   To warrant that the group succeeds, discipline is a must to be implemented by the members. It is the one quality of a group that can make it fail or succeed. Even talent, inspiration and good communication do not function without discipline in the group. The most disciplined teams are the most effective groups, and they usually have constancy of effort and great performance. The leader must lead by example by being disciplined.

8. **Eighth requirement for the virtue of group consciousness: Dispassion and detachment**

   When a group becomes fanatical, excessive or very emotional, the team can't function properly. In this case, the virtues of dispassion and detachment are needed. The members should not focus on the personal likes and dislikes and should subordinate their individual preferences to the good of the group. Any disturbing and disrupting agenda brought by anger, jealousy, envy, violence fear, seperativeness and stress should be minimized or dispelled immediately by the group or the leader. These negative energies should not be allowed to stay and grow in the group. The role of the leader and senior members of the group is to dismantle these poisonous qualities immediately. The training of the group members should include how to switch to the mental levels and exercise detachment whenever there are stressful and critical emotional situations. Being in the mental state, focusing on the next step and keeping the group mind focused on the long-term goal is an important protocol to develop detachment and dispassion.

9. **Ninth requirement for the virtue of group consciousness: Group love and compassion**

   Each member should be considered as the other member's keeper. There should be friendship and love for the other members, but not passionate or sentimental love. Why? Because too much emotionalism imbalances the group rather than binds it. The loving relationships should be colored with professionalism and integrity, and at times impersonality in relationships should be practiced if required. The more mental and advanced members can do this more easily than the rest. But it can be practical as a future requirement of more advanced group work. The true virtue of group consciousness is starting to develop in some groups and business organizations, but most groups and teams are still too emotional, sentimental and sometimes fanatical.

The requirements for the virtue of group consciousness just mentioned are simplified versions, but they can probably help some groups and their leaders who are still using a dictatorship style. A new process of co-creativity and conscious teamwork has to be developed. Other groups need to balance their excessive devotionalism with a more mental attitude and objective aptitude. Somc spiritual groups need to incorporate a more practical quality of group work, and many materially- and commercially-oriented teams or groups have to taper their excessive money-oriented schemes to be guided by more environmentally-friendly and spiritual goals. Whether you are a businessperson, educator, politician or religious leader, the virtue of group consciousness is a necessary ingredient to sustain power and achieve bigger goals in less time. All the spiritual values and virtues discussed earlier are required to master the virtue of group consciousness. It is one of the last virtues to be mastered, but once developed, you become unstoppable even in your path of spiritual evolution.

Seventh virtue: Good health

The key to attaining and enjoying lasting success and fulfillment

I have not seen a book that includes good health as a virtue. Therefore, I am obliged to declare good health a virtue and part of inner power development. All the other virtues are not easy to master and execute externally without good health. When you are sickly, will-power and vitality are more difficult to maintain even though you have the techniques to master them.

If a person has cancer and is in pain, it is difficult to cooperate in a team or to exercise discipline. It seems that when people are sick, the mental and emotional power cease to function properly. Even spiritual people who are very sick cannot perform their service and rituals and handle *kundalini* awakening safely. Talents and educational degrees are of no use when people are bedridden or suffering with life-threatening diseases, so it should be clear that good health is one of the most important virtues. A wealthy person will understand their powerlessness when they are between life and death. Their titles, success and money are meaningless in the end. If you have undergone a near-death experience, you will understand what I mean.

Whether we are parents, business leaders, politicians, doctors, spiritual leaders or multibillionaires, we need to respect the law of health and consider the virtue of good health as an important aspect of inner power. Most success and leadership seminars focus too much on strategies for business skills and emotional and mental development, but do not address the maintenance of good health enough as a most required attitude for success. Humanity has not yet realized this since vices that destroy health are still acceptable in even acclaimed leaders and brilliant people.

For example, I know a super-wealthy and successful business leader, who had a stroke at age 55 because of overwork, stress and vices and was bedridden with paralysis for many years. He said to me that his billions didn’t mean much to him after that. I have had many successful wealthy people as clients who were the best in their field, but one day lost their health and had to stay in bed for the rest of their lives. Therefore, I can't over emphasize the need to take care of your health. Even medical doctors are exposed to very stressful lives and work, and many of them still smoke and drink, and some of our well-known scientists and brilliant leaders are not the best examples for this virtue of good health. In fact, good health is not a priority of modern education as much as fame, money or career. No wonder many people don't take it seriously until they get really sick, which is sometimes too late. In the future, we will see that good health will be treasured and considered a priceless gift equal to professional success. Good health will be a demonstration and barometer of spiritual maturity and a high culture.

How do we enhance our good health?

Good health is not only the absence of physical ailments or discomfort, but it is complete physical, emotional, mental and spiritual well-being. It is wholeness and balance of all the aspects of a human being. Since there is no one single expert on all the levels for now, we need to itemize levels of care of human bodies and consult different experts.

Care of human bodies: Physical level

Today physicians are still the most reliable experts with regard to the cure of physical diseases, although there an increasing number of new alternative medicine professionals provide care for the physical body, including prevention of unhealthy conditions. Here are some general choices of currently available physical body care:

a. Consult your physician, dentist, psychologist and healers periodically.
b. For body structure and alignment problems, consult chiropractors and osteopaths.

c. Try massage and other body work for combined physical and vitality work
d. Experiment as needed with other alternative health care like Oriental medicine, ayurveyda and homeopathy.
e. Consult with wholistic nutrionists and other experts for a good diet. Spiritual teachers recommend a vegetarian or semi-vegetarian diet for advanced spiritual practices.
f. For physical body fitness training, get a fitness coach or consultant. Martial arts training, especially internal martial arts like *tai chi*, *chi-gong* or *aikido*, are more for energetic than just physical training

Depending on your physical body condition and requirements, you may choose from the above recommended body maintenance methods. It is still recommended for each individual to be responsible for managing their health by:

a. Sufficient rest and relaxation
b. Recreation
c. Proper diet and nutrition
d. Removal of all vices. Consult experts on cessation of smoking, alcoholism, and recreational drug use. If possible, avoid prolonged exposure to people with vices.
e. Proper exercise and breathing techniques
f. Proper physical hygiene, including a clean physical environment

Care of human bodies: Etheric and vitality health

Vitality is the fuel of your physical body, therefore it directly affects your physical health. By maintaining high vitality constantly, your ability to overcome stress and disease is better.

Recommended practices for etheric & vitality health

1. MDP Body Therapy and Inner Power Development Program (3 times weekly)
2. MDP Synchronized Breathing Technique (daily or 3 times weekly)
3. Combination breathing (3 times weekly) or slow abdominal breathing (daily)
4. Make sure you get enough fresh clean air daily and, as much as possible, a few minutes of sun in the morning, which will help revitalize your energy body. If possible, ventilate your house or work space and expose them to sunlight.

Care of human bodies: Psychological health

This is a combination of emotional and mental health and psychic conditions. We are living in an ocean of emotional and mental vibration, therefore this psychic environment also affects our health. The people we work with and those who live with us also affect our emotional and mental health. Therefore, it is more complicated than just keeping ourselves emotionally and mentally clean. The city we live in is also stressed and polluted with many types of contamination, including negative psychic energies.

How do you maintain your emotional and mental health?

Techniques to maintain emotional and mental health

1. Maintain high vitality constantly through exercise, breathing techniques, sufficient rest and proper food. It is easier to overcome all the surrounding pollution if your energy is high and your stamina is better.

2. Release stress and maintain a balanced emotional and mental positive attitude. Today, stress is so prevalent in our modern life that we are getting used to it. It is affecting our health negatively whether we notice the side effects or not. Stress weakens the immune system, drains physical, emotional and mental energy and lowers productivity and performance. Therefore, it is one of the weaknesses in most people, even the most successful people. Before stress overwhelms you, let's correct or prevent it immediately:

   a. For quick release, use the MDP Synchronized Breathing Technique for 5 minutes daily.

   b. For long-term solutions and to remove the side effects in the physical body, do the MDP Body Therapy and Inner Power Development Program for 20-30 minutes, 3 times weekly.

3. Heal your emotions and your mind for inner peace and to remove fears, phobias and negative psychological influences. Do the MDP Inner Renewal Meditation, 3 times weekly (see page 395).

4. Purify your psychic space by burning sandlewood incense. If you observe almost all religions, they use incense to purify their temples or worship areas. Even yogis use incense, especially sandlewood, to energetically clean stress and negative vibrations from their meditation room. You may also use the *Chantrams of Transformation*, a CD with special sacred chants that transform the energy of the listener and the environment.

5. Equip your emotions and mind with more advanced faculties, and continuously study and engage in social and mental activities. Keep the emotions and mind alive and active. Use them for service. Be futuristic.

Care of human bodies: Spiritual health

This aspect of life can be enhanced by practicing your religion or other spiritual sciences and techniques properly. This book offers some practical methods for improving character, purifying energy bodies and chakras, enhancing inner faculties and virtues, and gives you a new approach to deliberate spiritual growth. Doing service for a good cause helping human development and contributing to world evolution is a must if you want to accelerate your spiritual development.

To maintain your spiritual health as a way of life, you need to include:

1. Continuous study of universal principles and spiritual subjects
2. Applied service corresponding to our spiritual development
3. Meditation and spiritual practices as a source of inner connections and power
4. Proper development and use of faculties and inner powers for the greater good

5. Participation through group work for human and world development. The more advanced a person becomes, the more spiritual health encompasses and integrates all the other aspects of health. In fact, at higher levels of inner training, physical, etheric, emotional and mental health are subordinated to spiritual goals and purpose.

The virtue of good health will increasingly become a priority of the incoming culture because it will be the gauge of the quality of life one has lived, and good health will be a demonstration of inner development and balance.

Eighth virtue: Virtue of sacrifice

The key to great achievements and powerful service and the power to become a legend

In sports, politics, business, family or any endeavor, the virtue of sacrifice stands out as one of the keys to achieving great things and expressing a powerful life of service. Big sacrifices lie behind every great success: sacrifices of time, energy, effort, hard work, health and sometimes one's personal preferences. Great spiritual figures in world history like Jesus the Christ had to sacrifice his life to embody the spiritual principle of love. The Buddha had to sacrifice his comfort as a prince to become an ascetic in order to discern the virtue of wisdom. Martin Luther King's life was relinquished for the sake of civil rights. All the great leaders, heroes and saints are masters of this virtue. To leave a legacy, you need this virtue. It is the bottom line of greatness. The best chess players are those who understand the principle of sacrifice in the early stage of the game in order to win at the end. This principle is partially stated by Sidney Howard:

One-half of knowing what you want is knowing
what you must give up before you get it.
- Sidney Howard, American dramatist

The team that has a better chance of winning exerts extra miles of effort and energy in preparation compared to other teams. The principle of sacrifice is not what most people think. It is not relinquishing something bigger or greater for nothing or lesser. True sacrifice should not be seen as a trigger to suffering. In fact, it is a cause of joy or bliss. It is usually giving up a penny for a hundred dollars. Sacrifices almost always deserve greater rewards of contentment, joy and a feeling of greatness.

How can one start to relinquish lesser preoccupations in order to achieve greater things? The letting go of less important stuff – material, emotional, mental and even so-called "status" and beliefs for the sake of greater goals and causes. Giving up unproductive activities to do more important work and service that contributes to society or world need is a noble idea that older Souls do out of habit. But sacrifice is it is not martyrdom. It is selflessness and service-orientedness, but intelligently applied. It can be systematically implemented as follows:

Techniques to intelligently apply the virtue of sacrifice

1. Determine and list your most important goals in life that are the highest priority and the amount of time, resources, effort and attention you need to make them very successful yearly or monthly.

2. List all the less important goals and activities that can be reduced, relinquished or terminated in order to have more

time, resources, energy and focus for your priorities. Implement the screening process immediately.

3. Determine your physical, emotional and mental attitudes and aptitudes that block or disrupt your effectiveness to achieve the important goals. Implement the changes, adjustments or reversals immediately.

4. Identify your personal desires and preferences that need to be relinquished or changed because they are not supportive of your new priorities. This may include relationships or passions for a thing in the past that has sentimental value. Great personal sacrifices require courage, determination and commitment. This is where many people fail. They are very attached to the past, temporary possessions or relationships.

5. The ultimate reason, greater good and benefits of your goal – personal or for the good of humanity – constantly have to be remembered and focused upon to justify the reason for sacrifice. This is important for the individual or team.

6. People who do not have this virtue yet should start practicing by sacrificing smaller items within their capabilities. As it becomes a habit, then bigger things can be relinquished for greater goals.

7. The reward of being able to serve a great cause that contributes to life and society should justify the sacrifice, not just financial reward or fame. Many rewards supersede money and glory. These can include leaving a legacy and spiritual monument of greatness. Not all of the most remembered and revered beings of world history had publicity or money to

achieve greatness. Many have exemplified the virtue of sacrifice and service to humanity. The good news is that we can do it too and it is never too late. It is always good to bear in mind that mastery of this virtue of sacrifice is mastery of leadership and greatness. It is the power behind great achievements.

Practical meditation to enhance all the virtues

The MDP Inner Renewal meditation summarizes many of the applications and techniques to develop your faculties and virtues further. By practicing this meditation, your inner powers are enhanced rapidly, safely and naturally. I recommend that it be practiced regularly at least three times weekly as your basic spiritual practice or, if you do other practices for inner development, you may follow the weekly schedules recommended in Part III, page 429).

MDP Inner Renewal Meditation

(3 times weekly)

- For deep relaxation, peace of mind and stress management
- For inner empowerment and spiritual development
- For enhancing emotional and mental aptitudes and attitudes
- For purifying, unblocking and rejuvenating the different energy bodies
- For improving relationships and healing your past
- For activating the expression of virtues and to integrate the faculties of love, light and power in your life

Procedure

A. Proper Posture

1. For this meditation, the best posture is to sit in a lotus or semi-lotus position or on a chair with your feet flat on the floor and your hands in your lap. You may also lie down as long as you keep your back aligned.

2. Close your eyes.

B. Inner Purifications and Breathing

1. Start with the inner purifications. Free yourself from any physical, emotional and mental negativities or discomforts.

2. Imagine yourself in front of an ocean, the most beautiful ocean you have ever seen. What is important is your intention.

3. Now, breathe deeply and slowly and relax. During exhalation, have the intention to breathe out your negativities and discomforts.

4. Inhale deeply and breathe out to the ocean any discomfort, tension or pain in your body. Do it many times. Let the ocean absorb the discomforts and impurities. (1½ minutes)

5. Inhale deeply and breathe out to the ocean stress or emotional pain and discomfort. Let the ocean absorb these negativities. (1 minute)
6. Inhale deeply and breathe out anger or irritability. Let the ocean absorb the negative emotions. (30 seconds)
7. Inhale deeply and breathe out depression or boredom. Let the ocean absorb it. (30 seconds)
8. Inhale deeply and breathe out fears, guilt and worries. Let the ocean absorb the negative emotions. (30 seconds)
9. Inhale deeply and breathe out resentment and anxiety. Let the ocean absorb these negative emotions. (30 seconds)
10. Inhale deeply and breathe out any remaining negative feelings and emotions. Let the ocean absorb all of the negativity (30 seconds)
11. Inhale deeply and breathe out harmful thoughts against anyone or any being. Let the ocean absorb the negative thoughts. (30 seconds)
12. Inhale deeply and breathe out any mental limitations and blockages. Let the ocean absorb the negative thoughts. (30 seconds)
13. Inhale deeply and breathe out any tendencies of excessive criticism, control and manipulation of others. Let the ocean absorb all of the negative tendencies. (30 seconds)
14. Inhale deeply and breathe out anything that limits your freedom or obstructs your inner development. Let the ocean absorb any obstacles. (30 seconds)

15. Relax and enjoy your new emotional and mental well-being.

16. Imagine the beautiful ocean gradually becoming a mixture of pink and golden light, a brilliant pink and golden light. Mental intention is all that is required.

17. Walk towards the ocean of light. Immerse your entire being with pink and golden light. Feel and be aware of the soft, silky sensations all over your body. Inhale deeply and exhale slowly and allow yourself to absorb & assimilate the fresh, revitalizing energy. Take your time. (30 seconds)

C. Opening and activating the heart center

1. Now, be aware of a point in the center of your chest. This is an area where you can experience love, inner peace and inner joy. Inhale deeply and exhale slowly while being aware of the area in the center of your chest. (1 minute)

2. As you breathe, silently affirm:

   I am a peaceful, loving and joyous person. I am a peaceful, loving and joyous person.

   Repeat it many times as you continue breathing slowly. Create this wonderful feeling of love, joy and peace within you. Express your love, gratitude and respect to those who have helped you in your life.

3. Visualize yourself being more loving and caring to other people and their needs. Start with your family and loved ones. Express love to your friends. Express good wishes & respect to your co-workers. Mentally ask for forgiveness

from people who have been hurt by you. Now, forgive the people who have hurt you. Experience the love, peace and joy within you.

D. Activating your mental intelligence and will

1. Now, focus on a point between your eyebrows and breathe slowly. By breathing while focusing on this point, you can enhance the ability to think more clearly and properly. Inhale deeply and exhale slowly as you mentally say 3 times:

   I am a creative and intelligent person. I will use my creativity and intelligence lovingly and properly.

E. Opening and activating the crown chakra for spiritual powers and intelligence and aligning to the Soul

1. Now, focus on top of your head and breathe slowly. By breathing while focusing on this point, you can enhance the ability to acquire more inspiration, Divine understanding and wisdom. Inhale and exhale slowly and mentally say 3 times:

   I respect the Divinity within myself. I respect the Divinity within All. I respect the greatness of the Supreme Source of Life.

2. Relax and let go. Enjoy the inner peace and stillness. (2 minutes)

F. Absorption and circulation of internal energy

1. Now, to assimilate the internal energy & vitality, breathe slowly while focusing on your navel. Do this many times for 1 minute.

2. To further improve the absorption of natural energy, concentrate on your palms & soles of the feet as you inhale and exhale slowly for 30 seconds.

G. Blessing service (2 minutes)

1. Now, with intention, share some of your positive and loving energy & vitality with your loved ones, friends and anyone who needs help. You may use the following affirmation:

   I humbly offer myself as an instrument to share light, love, peace, joy, beauty, truth and goodwill to all of humanity and to the whole Earth. So it is!

Now open your eyes and smile. How do you feel?

The last method to be learned in this book is the technique of Spiritual Blessing. By incorporating the power of the Soul into your incarnated personality, you can become a potent instrument or channel to share spiritual blessings of love, peace, joy, harmony and healing energy. This is called the MDP Soul Radiatory Blessings.

MDP Soul Radiatory Blessings: Applications and Service

- To spread the energy of love, peace, joy and harmony to individuals, your family, groups, organizations, nations or the Earth
- To bring healing and transformation to others and the Earth
- This technique automatically develops certain inner powers, especially through the virtues of benevolence, altruism, group consciousness and good health. Your whole human energy system will be purified and revitalized as you perform these techniques.
- This is individual spiritual service at a global level in just a matter of minutes of blessing work.
- Do not use this technique if you have an addiction to smoking, alcohol and/or recreational drugs or if you have life-threatening conditions. The coarser and negative energies can be transferred to others when you have these conditions. It is also not recommended for people with heart ailments or high blood pressure.

Procedure

(18-20 minutes, 2 times weekly)

1. Do the MDP Synchronized Breathing Technique. (3 sets)

2. Focus on your heart, ajna and crown chakras on all 3 levels as much as possible: etheric, emotional and mental. Breathe slowly for 1 minute.

3. While focusing on the 3 chakras, know that you are enveloped and inside your Soul, just like being in a bright light or Sun-like radiance. Continue this awareness as you breathe slowly.

4. Visualize your subject of blessings, be it an individual, your family or a group. Here is a suggested sequence to radiate Divine blessings for healing or transformation:

    a. Individual: adults for 30 seconds and younger children for 5 seconds. Do not do this for babies or infants.
    b. Family as a group: 1-2 minutes of blessing
    c. Your group of friends, associates or co-workers: 1-2 minutes
    d. Your organization: 2 minutes
    e. Your country: 3 minutes
    f. Humanity as a whole: 3 minutes
    g. Earth as a being: 3 minutes

How do you radiate Divine blessings? Just exude the energy blessings to the subjects, knowing that you are the Soul as in step 3. You are like a brilliant sun radiating and transferring light and love to your subjects. Visualize the subjects as a miniature in front of you. For example, imagine your family as small people in a group in front of you. Your country is shrunk to a small map, and humanity is imagined as a small ball of light in front of you. The Earth can be imagined as a very small globe. The way you radiate is very important because the power of these blessings lies in the concept that your light as a being is extremely brighter and bigger than your targets or subjects. You are like the sun, and the energy subject is a smaller size.

Your Soul Divine Light has extreme potency to purify, revitalize and transform the negativity of the energy bodies and chakras of individuals or groups. When you do it to countries, humanity or the Earth, your Soul's power also attracts and triggers invisible healing beings to do the energy transformation and blessings with you. Thus the real power is not only from you, but from a team of beings, so this service is group work. The ability, capacity and entitlement for you to be protected and endowed with good things in life are proportional to how much you give back to the earth, your country, humanity, your group, your family and to life. You can do this with your good intentions through this method's spiritual blessings.

> The quality of thoughts, emotions and blessings that you radiate is what you receive back. Passion begets passion, anger begets anger, Divine Love begets unconditional love and spiritual blessings attract
> Divine blessings.

You may wonder, why are many pictures of Saints, prophets and Sages portrayed and photographed with a posture of blessings using either their hands or their brilliant auras? It is one of their roles and duties in life to generously and benevolently share their Divine light, love and power with all who need these Divine attributes and qualities. As you do what they do unconditionally, you also tread their path of wisdom, service and life. I'd like to leave you with this simple reminder:

> Service is an expression of inner growth and development.
> It is the muscle-building for virtues.
> It is also a barometer of inner development and spirituality.

We have studied many principles through spiritual inner powers and the concepts and methods by which they are acquired or achieved. This section alone contains a lesson for a lifetime, but it is important to know that great world evolution has been achieved little by little, and these constant small steps have endured the test of time. The same with your personal evolution – the positive change can be slow, step-by-step. But with these new organized strategies, why not develop faster and integrate different techniques instead of a linear, step-by-step approach? Let's change the old concept of linearity. Let's be synchronized. Master as many techniques as possible in the shortest time.

How can these techniques and principles be part of your life? We'll make it simple and straightforward:

1. First, you're halfway there with the concept and technical knowledge of inner powers and the methodology to develop them. A strategy has already been provided; it's yours to implement.

2. Revisit the section on spiritual powers, and identify what aspects are your priorities and immediate important needs that make a big difference in your life. The principles were discussed in the most practical way possible as systematic procedures and techniques, which you should implement as a weekly practice. Master one or two virtues simultaneously every month and continuously roll them over every year. Start with the virtues that you need the most.

3. List in your planner or calendar your immediate priorities and create a realistic schedule for your chosen practices with dates, times and rhythm to do them, or you can follow the general schedules in Part III, page 429. This is part of implementing the virtue of discipline and constancy.

4. Commit to practicing them regularly. Proficiency comes with practice, practice, practice. Results are inevitable.

5. It is also advisable to gradually introduce these spiritual principles to your family members, friends, associates or organizations. Start with receptive people.

Reminders about inner powers

It is important to emphasize the following reminders:

1. The true esoteric mastery of inner powers is directly proportional to and demonstrated by the mastery of virtues, including that of good health.

2. The virtues are the result of applying and using the faculties, and each virtue is a key to unleash certain inner powers.

3. The permanence of the influence exemplified by the greatest human beings and Sages on Earth is directly proportional to the inner powers they utilize for their service to humanity and the world. This also applies to your inner power development.

4. All the external manifestations of individual or group power are but a reflection of the inner powers they posses or have developed.

5. The true sources of power are the Soul and Spirit, and there are great invisible forces and Beings that are the macrocosmic source of the powers of evolution on Earth.

6. Almost anyone can develop inner powers with systematic training and the right time is now.

Now that you have studied the different inner powers and how to unleash and develop them, maximizing your performance in your personal career and spiritual life is simpler and more defined. I hope that building your inner powers and applying them results in more balanced material and spiritual success in your life.

Part III
The Importance & Relationship of Inner Power to Evolution, Quality of Life & Practical Performance

Maximizing Performance.. page 410

Strategies for Maximizing Performance.. page 412

Recommended Weekly Programs for
Inner Power Development.. page 424

Weekly Schedules for Maximizing Performance &
Inner Power Development.. page 429

Part III
The Importance & Relationship of Inner Powers to Evolution, Quality of Life & Practical Performance

Maximizing your performance in both material and spiritual life is a good idea. Through inner power development, success and fulfillment are inevitable. By regularly applying the principles and practical tools in this book, a breakthrough in your life can lead you to physicalize your vision, mission and plans faster and more effectively.

Acquiring and developing inner powers is a good investment and lifetime pursuit for anybody who wants to be ahead. This is especially so for leaders, executives and professionals who want to stay on top with greater usefulness and capacity for service and power. The ability to achieve goals and produce excellent results faster is directly proportional to the inner powers developed, which include:

- Constant vitality and will-power
- Emotional intelligence and right human relations
- Integration of the concrete and abstract mental intelligence
- Insightfulness and visionary faculties through psychic powers
- Spiritual powers of wisdom, intuition and inspiration
- Integration of the above powers and faculties for spiritual and material balance, guided by the power of inner virtues and universality

Your individual, family and team mental powers, emotional intelligence and dynamic vitality are becoming most vital assets, especially in the fast-paced life of the modern time. The world and humanity are at a crossroads to move into explosive growth and a better quality of life at personal, family, group, national and global levels. This will entail not only technological advancement, but also a scientific approach to developing inner powers latent in humanity, especially in leaders and thinkers. Humanity has never been in such great search of new practical guidelines and tools to meet complex modern global problems and challenges. And this requires an inevitable need to upgrade personal aptitude and attitude in order to cope successfully with the speed at which the world is changing.

Technical skills and talents, formal education and material resources are not enough to maintain inner stability, personal security and peace of mind. Past formal education, information and career training serve less and less as a guide for present and future decision-making and problem-solving. Even world leaders and experts in different fields are expressing the need to redefine their visions and goals to give them farsighted global meaning and usefulness. Our planetary life has become unpredictable and complex, but it's complex only to those who haven't been exposed to the inner realities and powers that shape who we are and the world we live in. It is the unknowns and missing ingredients of truth that make life complex. This book seeks to make some of the unknowns known. Volumes II and III of *Inner Powers* shed even more light on these unknowns to further reveal the inner world of causes to humanity in order to give a new, higher perspective that recognizes individual as well as global challenges as opportunities to learn to become wiser and to evolve further.

Humanity needs new ideas, new solutions and new strategies through additional acquired inner powers. These will allow us to go through any challenges, even during the most difficult times, and to convert these experiences into an opportunity to maximize growth and inner development. The teachings in this book offer integrated practices and pearls of wisdom, which have produced rapid advancement in the development of many human beings and are available here for your practical use.

When you finish reading this book, revisit the lists of topics in each section and choose which aspects and qualities you need to develop first. Practice and apply the techniques immediately according to the recommended schedules.

So what now? How can you practically apply what you've learned to maximize your performance in life and achieve even greater results, success and fulfillment?

Maximizing Performance

When people speak of "great performance," it is usually based on how successfully positive goals or planned activities have been implemented and achieved. Performance is not necessarily the number of activities accomplished, but the quality of results based on the original targets or objectives.

In measuring performance, there is a need to look at the rate of output and input, as well as the period of time to accomplish the tasks or desired outcomes. A traditional way of looking at maximizing performance can be expressed in an equation:

> Maximum performance =
>
> Maximum output with minimum resources
>
> accomplished in the shortest time

This considers only quantitative and qualitative results without taking into account the following effects:

1. Immediate environmental, social or global effects of goals and their outcomes
2. Negative side effects on a workforce, team or individuals working on and
   implementing the goal and activities
3. How competitors and other individuals and parties may be affected by the results
4. Long-term effects, either on the individual, family, organization, nation, humanity or the world

Therefore, a better way of expressing the performance equation is:

> **Maximum performance = Maximum quality outcomes using minimum resources accomplished in the shortest time and resulting in "win-win" situations**

The quality of maximum performance expresses its effects over time. The second formula takes into consideration the quality of results concerning all people, parties and environmental and social effects over time. To date, this type of performance has been more of an ideal than reality for most leaders, groups and their organizations. I don't know anyone who has achieved this type of approach 100% of the time. But with the new techniques and faculties available in this book to bring inner power into the equation, you can develop a strategy to unleash your or your team's maximum potential.

Strategies for maximizing performance

There are several important requirements in order to achieve a high rate of success in any chosen path in life. Let's make them as universal as possible so the strategy is applicable to all types of people, cultures, careers and walks of life.

Strategies for maximizing performance
1. Maintain your good health: physical, vitality, emotional, mental and spiritual
2. Develop the highest quality proper aptitude and right attitude: • Emotional faculties and powers • Mental faculties and powers • Psychic faculties and powers • Spiritual faculties and powers • Integration of the above faculties and powers
3. Intelligent hard work to materialize your purpose and goals in life effectively

Let's discuss and summarize each component of this strategy.

First strategy: Maintain your good health at all levels

Health is wealth, and the one who has good health already has initial abundance and great potential. Those with big potential can physicalize many great realities. It is best expressed by this statement:

> He who has good health has hope and
> he who has hope can do almost anything.
> - Arabian Proverb

The purpose of maintaining your good health is not just to achieve your goals, but mostly to be able to enjoy your success after you have achieved it and to be able to cherish the fruits of your labor. It is good to be very successful, alive and especially healthy! Refer to Section E, Spiritual Powers, Virtue of Good Health, pages 386-392 for complete guidelines on:

- Care and management of physical health and vitality (page 387)
- Care and management of emotional health (page 389)
- Care and management of mental health (page 390)
- Care and management of spiritual health (page 391)

Second strategy: Develop the highest quality proper aptitude and right attitude

Developing the proper aptitude and right attitude to the fullest is your primary task to maximize your achievement and success. We cannot expect a toothpick to support a table. Likewise, we cannot expect a mediocre person to manage an empire or lead a global project. What makes a great, accomplished person different from the ordinary person is aptitude and attitude. Extraordinary achievement in life is only possible with the proper aptitude and right attitude. To be able to live with greatness and nobility, one must master aptitude and attitude. What are the distinctions between aptitude and attitude?

1. **Proper aptitude**: Proper attitude relates to physical, emotional and mental technical skills and talents, whether they pertain to a mechanical skill or to leadership acumen. Aptitude is developed through education, continuous training and experience. At the higher levels, it includes the ability to manage personnel, resources and investments. It also includes the ability to plan and produce results effectively.

2. **Right attitude**: This is an expression of emotional, mental and spiritual behaviors and psychological attributes brought by spiritual

development and personal maturity. Right attitude is not only useful to attain a highly cultured personality, but is also a necessary quality to bring you from success to fulfillment. How you perceive and appreciate what you have in life and fully sense the greatness of life itself all depends on your attitude. It is related to your subjective existence and inner nature.

A person can have the proper aptitude, like good technical management skills, but not necessarily have the right attitude. One can be wealthy and materially successful with the right skills and good aptitude at work, but lack the right attitude to maintain good family or social relationships. On the other hand, many spiritual individuals have the right attitude of benevolence, but do not have the proper skills and aptitude to make larger contributions to society. Both attitude and aptitude have to be developed, cultured and integrated to maximize your performance materially and spiritually.

> An individual who has the proper aptitude can be the technical leader and the one who has the right attitude is the inspirational leader, but the person who has both the proper aptitude and right attitude is the leader of both.

This book offers many sections of practical applications and ways of culturing your attitude and cultivating your aptitude. Aside from technical training in your career and information from traditional education, I recommend that you revisit the different sections of Part II and select the techniques and topics you need immediately. This book is like a library of principles and teachings; just take what you need most first and then go to the other less important subjects you need to master later. This book generally provides most people with the required emotional, mental and spiritual training for a lifetime of practice.

Let's review aptitudes and methods for their development.

Methods for aptitude development

1. **Emotional power and intelligence** (Part II, Section B, page 89)

   This can help you with the following:

   a. Development of the right passion for the cultivation of courage, determination and commitment (page 104)

   b. Cultivation of the power of the heart for right human relationships, love, compassion and aspiration (page 104)

   c. Managing and releasing your stress daily and overcoming your emotional limitations (page 97)

   d. Development of a positive and powerful emotional aura of attractiveness and inner beauty (page 108)

   e. Development of emotional and social intelligence through an attractive energetic presence (page 106)

2. **Mental power and intelligence** (Part II, Section C, page 116):

   The following techniques and topics help you develop a powerful mental aptitude and a highly cultured mind:

   a. The power of common sense for reasoning ability (page 130)

   b. The power of a high grade of common sense (page 134)

   c. The power of decision-making and problem-solving (page 140)

d. The power of research and investigation (page 142)

e. The power to concretize or physicalize ideas and projects (page 143)

f. The power to organize life and manage time (page 144)

g. The power of innovation and creativity (page 154)

h. The power of imagination to harness the power of creative ideas (page 157)

i. The power of the love-intellect faculty for balanced problem-solving and decision-making (page 173)

3. **Psychic powers** (Part II, Section D, page 181)

Developing these inner powers and aptitude will give you extraordinary performance beyond the reach of most people today. It will allow you to attain a cutting edge aptitude with winning edge capabilities in your chosen career or line of duty. The additional psychic power development and faculties will help you:

a. Improve decision-making and problem-solving skills with more accurate instincts and feats that can sense beyond your mind and common sense
b. Enhance your ability to predict or forecast the incoming possibilities of your life and goals
c. Develop your ability to access higher ideas and knowledge for personal growth, professional success and greater service
d. Improve your sensitivity to be aware of and experience the subjective and inner world consciously

In the future, there will be a different type of blindness, deafness and numbness in human beings. It will be the inability to see, hear, feel and sense the subjective reality and inner world. Currently, to be a cultured person, you have to observe social codes of conduct and right etiquette. In the future, the real culture of civilization will transcend physical displays of behavior. It will include the ability to behave properly in both the outer and inner worlds, and this will be possible with the development of the psychic powers. With the development of psychic powers, you can further extend your service and exploration into realms unknown to most people. Therefore, you can become a pioneer with an extraordinary gift of inner knowing.

4. **Spiritual powers** (Part II, Section E, page 290)

The faculties and attitudes you gain through spiritual inner powers will make you totally effective and able to perform better to sustain success for as long as you want. There is nothing more important to allow you to enjoy your success and transmute it into the higher level of fulfillment than the spiritual powers, especially the power of virtues. Here is a review of the spiritual powers, virtues and their sources:

a. **Power of enlightenment** (page 295)

This is a state of accomplishment that allows you to harness the power of the Soul to fertilize and polish the aptitude and attitude of your personality

b. **Power of the Earth's energy planes, Invisible Spiritual Forces and groups** (page 310)

This provides you with power and development beyond your limited personality. To tap into these beings and their dimensions is to connect to an unlimited reservoir of inner power – to be used for greater service and a higher purpose

c. **Power of the Earth's sacred fire or *kundalini*** (page 320)

This is the creative force that stimulates all growth in humans, invisible beings and even the Earth as an alive Being.

d. **Power of sacred words and universal sounds** to align yourself to the universal source of power and energy (page 333)

e. **Power of virtues** to bring out the best culture in your personality (page 338). Here is a summary of virtues and their significance:

- **Benevolence**: The power to achieve happiness, contentment and right human relations and the key to accessing greater inner power
- **Altruism**: The key and power behind prosperity and sustained abundance
- **Discipline and constancy**: The key to attaining greatness and achieving big results
- **Objectivity and practicality**: The power of discrimination to achieve balanced results – the antidote to fanaticism
- **Will-power and vitality**: The key to speed in achieving results and continuity of long-lasting performance
- **Group consciousness**: The big key to sustaining power and achieving greater goals in less time
- **Good health**: The key to attaining and enjoying lasting success and fulfillment
- **Sacrifice**: The key to great achievement and powerful service and the power to become a legend

f. **Power of meditation** (page 395)

The meditation techniques in this book are universal and can be used as a tool for practical spirituality. The advanced meditations we included are considered as tools and processes rather than the goal. This is where many practitioners are a little bit misled. They make meditation the end goal rather than a step to a higher goal. There are many kinds and styles of meditation taught nowadays, but most are mystical in nature, making you relaxed or spacey. Of course, one of the most important purposes of advanced meditation is to achieve spiritual oneness and connectedness to the Soul or Spirit, but there are other practical uses like purifying and revitalizing your energy bodies, healing yourself and important activation of your different chakras to stimulate their corresponding faculties and powers. One of the most important goals in meditation is to awaken the sacred fire of *kundalini* for rapid inner development. All of these aspects of meditation are included in the MDP Inner Renewal meditation, which can help you grow personally and spiritually.

g. **Power and service of blessings** (page 401)

The power of blessing and transforming others is one of the most noble services you can perform. This is a spiritual work that most, if not all, Saints and Sages do daily – to share their blessings of love, joy, peace, harmony and healing energy with as many people as possible. As you bless others abundantly, you are also blessed with good things in life. This service fulfills the universal Law of Cause and Effect. The good luck or invisible abundance generated through your daily blessings can be received back as good health, Divine protection, guidance and great fortune.

If you want to always be on the right track in life with maximum benefit in everything you do, with all the aptitudes and attitudes you require to perform at your best, master the spiritual powers, especially the inner virtues. They are practical, with spiritual principles tied to inevitable results of productivity, right human relations and effectiveness in life. Your ability to achieve your maximum potential with fair win-win situations depends on the clarity of your conscience and mind – and this requires the development of the virtues. Every virtue you practice opens a portal of power in your life.

Third strategy: Intelligent hard work

Hard work is not enough. Creativity does not stand alone. Faith alone is not sufficient. Good karma without a way of expressing the precipitation of abundance is not effective either. So what is required?

> Intelligent hard work =
> Proper organization and management of your goals/project +
> disciplined work + good luck

Let's analyze it:

1. Properly organized and managed goals and projects

entail:

a. **Prioritizing goals and activities** depending on your life's vision and purpose. All of your roles, plans, targets and results should revolve around your vision, mission and purpose.

b. **Managing your life by managing your time**. We have explained this in Part II, Section C, page 144 with a technique: Power to organize life and manage time.

There are **five most important areas** you need to include in your life planning and time management:

- Family and home
- Career and work
- Social and environmental contribution
- Spiritual life
- Health and recreation

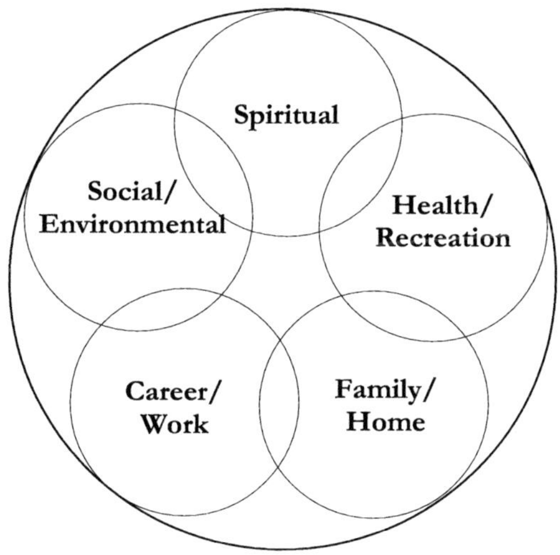

Integrated Life Planning

Figure 36

All of the activities and schedules you need to implement daily fall into these five areas. Balance your success by allocating a reasonable percentage of your time to satisfy your roles and goals according to these five most important areas of your life. This is the key to attaining not only success, but also fulfillment.

c. **Materializing your goals with a systematic method**. There are always better ways to achieve maximum results with minimum input in minimum time. Apply the method in Part II, Section C, page 143: Power to concretize and physicalize your goals and projects. It is a simplified technique that has worked for many executives and clients. Apply it and see the results for yourself.

d. Mastering the power of decision-making and problem-solving (page 140)

Very accomplished people and super-achievers share two basic qualities: good decision-making and problem-solving capabilities. Without these, a person cannot lead or possess great power.

2. Disciplined work:

Disciplined work does not mean overworking. It means allocating your time to only the most important areas and activities. The virtue of discipline discussed in Part II, Section E, Spiritual Powers (page 359) includes all of the requirements of this virtue. Revisit that section and master its application. Again, when one has discipline and constancy, success is predictable. Remember that the most successful people are those who are able to go the extra mile in effort and spend more time doing the most important things for greater value. Disciplined hard work requires the virtue of sacrifice. We also discussed this in Part II, Section E (page 392). Behind every great success in any chosen endeavor are great sacrifices – time, energy, effort, resources. To leave a legacy through your work, you need this virtue. It is the power behind big achievements and greatness. True disciplined work needs the power of the mind to be able to focus on the work at hand while maintaining a complete awareness of and adherence to the long-term vision. Always finish the goals and projects you start. Make discipline a way of life.

3. Good luck and entitlement:

We discussed this in Part II, Section C, Programming and materializing goals (page 145), but let's summarize the important points here. Ancient as well as current endowments of good karma, or what we call "good luck", entitle people to receive good things or allow them to materialize goals faster. This good luck is also a force that helps remove obstacles in life. We mentioned earlier that good luck and the ability to be entitled to good things go to the people who work for them. Also, the speed of achieving goals and removal of obstacles are directly proportional to the amount of karmic entitlement available to an individual or organization. I've discussed this topic and different related techniques in Part 2, Section E, Virtue of Altruism (page 349). If you want to develop the power to materialize goals fast and need to increase the speed of removing obstacles, you have to practice and master the virtues of benevolence and altruism. The good karma generated by these two virtues is so potent that in the absence of other virtues, people can physicalize the same goals faster and easier than those who work harder without the good luck to harvest good results.

I recommend several techniques and practices weekly to start the mastery of inner powers. By simply combining these methods properly, your performance will dramatically increase with in a few weeks of practice, your aptitude and attitude will be enhanced and the balance of your material and spiritual life will definitely come naturally. You might want to consider making these sequences of practices your basic foundation for inner power development. If you need to specialize in a specific cultivation of a certain power from any part of the book that is not mentioned in the basic schedule (see page 429), then add that technique to your weekly schedule.

Recommended weekly programs for inner power development

Goodwill Programming: 5 minutes daily anytime, but preferably immediately after you wake up and before you start your day

Procedure: Lying in bed or in a sitting position

Start your day with good intentions and express good wishes to as many people as possible. How you start your day determines the flow of events throughout the day.

a. **1 minute**: Express gratitude for your life and to the Source of Life from your heart.

b. **2 minutes**: Silently express your good intentions and wishes to your loved ones, family, friends, co-workers and people whom you will meet throughout the whole day. Visualize your relationships with everyone enhanced.

c. **2 minutes**: Visualize or form the intention that your whole day is very successful and that your goals for the day have been achieved. See yourself happy and harmonious at the end of the day. If you have personal problems, visualize them being resolved. But make sure that you always focus more on the positive outcomes.

This ritual can also be done before you sleep or anytime you need to enhance relationships with people. Refer to Part II, Section E, Virtue of Benevolence (page 339).

1. MDP Body Therapy and Inner Power Development Program

20-25 minutes (pages 54)

Goals:

- Rejuvenate and revitalize the body quickly
- Release stress and tension
- Enhance metabolism and accelerate weight loss
- Oxygenate the blood and detoxify the body
- Enhance your total health and well-being

This can be practiced daily or as per the basic weekly schedule.

2. MDP Synchronized Breathing Technique

5-10 minutes (page 81)

Goals:

- Quick stress management technique
- Purify and calm your emotions and mind and boost vitality quickly
- Rejuvenate the human energy system fast
- Stimulate Divine connection and increase Soul energy flow to the personality

This can be practiced daily or anytime you need to detoxify and revitalize your bodies. This breathing method is one of the most useful techniques supporting other practices.

3. MDP Inner Renewal Meditation

30-40 minutes (page 395)

Goals:

- Culture the different energy bodies
- Inner purification and revitalization of the human energy system
- Activation of the different inner powers of the mental, emotional and etheric bodies and improve the connection of the personality to the power of the Soul
- Healing of relationships and past negative tendencies
- Service through blessings

This meditation can be practiced anytime you need healing on any level.

4. Slow Abdominal Breathing: 5 minutes (page 77)

Goals:

- Increase your vitality and improve stamina
- Enhance concentration and mental focus
- Partly awakens the *kundalini*

This can be done anytime you need to boost your vitality. It can be performed in combination with the MDP Synchronized Breathing Technique.

5. Special Additional Practices (optional)

a. **Will-power enhancement**: Use any method that suits you. Refer to Part II, Section E, Will-power Enhancement Techniques (pages 376)

b. **Awakening the *kundalini***: Use 2 methods alternately: The simplified *mantric* yoga technique (Part II, Section E, pages 333) and the nostril controlled breathing technique (Part 2, Section E, pages 326)

c. **Psychic power development**: 20-30 minutes (Part II, Section D, pages 181)

Determine what psychic powers you want to master first and practice the corresponding technique for a month before you shift to a new psychic power development method. Rotate the different practices until you master all the available techniques.

d. **Pure thinking time**: Allocate at least 1-2 hours weekly to develop your mind, especially the abstract mental faculties, or practice the different mental power applications. Your thinking time should not only focus on your career. It can be higher study of new spiritual subjects that are of immediate interest for your next step. This is very important for mental muscle-building and for the requirement of integrating the concrete and abstract mental powers.

As you spend many months practicing and training, you might need to increase the frequency of the weekly schedules. Volumes II and III of *Inner Powers* will offer more advanced techniques and practices that might attract your attention and interest. The techniques of the three volumes have been designed with continuity, coherence and synthesis. Therefore, the practices built up from Volume I of *Inner Powers* will always serve as a foundation to be adapted to the more advanced spiritual technology of the other two volumes of *Inner Powers*.

SUMMARY OF SCHEDULES

Culturing the personality:

I. For normal training and inner power development

	Weekly schedule of allocated time (minutes)						
Practice	Mon/ Time	Tue/ Time	Wed/ Time	Thur/ Time	Fri/ Time	Sat/ Time	Sun/ Time
Goodwill Programming	AM 5 PM	5	5	5	5	5	5
Body Therapy	AM 25 PM			25			
Inner Renewal	AM PM					40	
Synchronized Breathing	AM PM 5		5		5		
Slow Abdominal Breathing	AM PM 5		5		5		
Kundalini Awakening: *Mantric* Yoga & Nostril-Controlled Breathing	AM PM		*Mantric Yoga* 20				Nostril-Controlled Breathing 15
Psychic Power Development	AM PM						30
Will-Power Development	AM PM	10					
Blessing Service	AM PM 10				10		
Purely thinking and study time	AM PM					60	
Total time invested weekly	AM 30 PM 20	15	5 30	30	5 20	105	50

Investment: AM 4 hours (morning schedules)
PM 1 hour, 10 minutes (evening schedules)

II. For healthier and more advanced practitioners

	Weekly schedule of allocated time (minutes)						
Practice	Mon/ Time	Tue/ Time	Wed/ Time	Thur/ Time	Fri/ Time	Sat/ Time	Sun/ Time
Goodwill Programming	AM 5 PM	5	5	5	5	5	5
Body Therapy	AM 25 PM		25		25		
Inner Renewal	AM PM	40		40			
Synchronized Breathing	AM PM 5		5		5		
Slow Abdominal Breathing	AM PM 5		5		5		
Kundalini Awakening: *Mantric* Yoga & Nostril-Controlled Breathing	AM PM	Nostril-Controlled Breathing 15		Nostril-Controlled Breathing 15			*Mantric Yoga* 20
Psychic Power Development	AM PM	30					30
Will-Power Development	AM PM 10		10				
Blessing Service	AM PM 10		10		10		
Pure thinking and study time	AM PM					2 hours	
Total time invested weekly	AM 30 PM 30	45 45	30 30	45 15	30 20	2 hours	55

Investment: AM 5 hours, 55 minutes (morning schedules)
PM 2 hour, 20 minutes (evening schedules)

The 21st century is a good time to shift to new frontiers of scientific discoveries and breakthroughs accompanied by unbelievable leaps of consciousness and spiritual realizations that were never thought possible. If you begin where you are now to apply these teachings on inner powers, you can improve your life and the lives of those around you tremendously. Humanity's success and fulfillment in the 21st century will depend primarily upon the extent to which we and our children develop and master the inner powers available to us. This will result in the long sought-after global unity, peace and balance of material and spiritual development for you and humanity.

The spiritual principles and inner power development methods in this book are similar to the laws of physics or mathematics; when you use them, the corresponding results are foreseeable and measurable by your increased productivity at work and the change of your perspective in life. Constant vitality and stamina will be yours, discipline and constancy will supersede inconsistency and life's ups and downs, inner peace will replace stress, daily effectiveness will replace periodic efficiency and inner power and development will shine over uncertainty and bewilderment. The formula and strategy are yours to apply. Some of these methods may seem simplistic and basic in approach and sometimes appear to repeat themselves in different ways. However, they're not exactly the same applications when combined with other steps. It is the right integration and synergy of the different teachings that makes the results powerful.

Expertise in any field always goes to those who master the basics. Therefore, it is recommended that the fundamentals in this book be understood and practiced with constancy.

Volume I of the *Inner Powers, Maximizing Performance Through Inner Power Development*, has been geared to helping you develop your inner powers in life so that you can become the best you can be in the shortest time possible. If you follow the steps and techniques in this book, you can enjoy a lifetime of maximized performance, lasting success and fulfillment without sacrificing your personal life and service to society and the world. Now that you have started your inner journey with certain knowledge of inner power, it is my hope that you have been inspired to understand that behind every outer expression of power lies a related inner power, which is the real source of its continuity and persistence. Being aware of and connecting to these sources – your Soul, your Spirit and greater beings of evolution – and their superior qualities can give you a different perception of yourself, your life, the world around you and your place in it. And this gives you a nobler purpose and higher vision to expand your consciousness and effectiveness as a human being – whatever your chosen career and path – to heights you may never have imagined possible.

The extent of your greatness and performance is limited only by your perception. Just reading this book has opened many new windows to higher perception and truths. With regular practice, you can open not only more windows to higher truth, but also bigger doors to greater consciousness. As you move further into a world of universal principles and causes – guided by inner powers – you will inevitably build a lasting legacy of service to your family, profession, community, humanity and the world. Consider this:

Think benevolently;
Speak wisely;
Act fairly;
Work intelligently;
Share altruistically;
Serve selflessly;
Live moderately;
Success follows…
Fulfillment thereafter.

Your journey has begun. Best wishes for your continued success and fulfillment.

GLOSSARY

Adi: The Earth's supreme plane composed of 7 layers of energy frequencies. This plane is the highest and most refined state of consciousness in the planet.

Ageless Wisdom: A compilation of esoteric science and inner spiritual teachings that have been handed down from ancient spiritual sources through Sages and Masters of wisdom

Anjali mudra: A Sanskrit term referring to a hand position used by yogis, monks or meditators that joins the hands at the chest level with fingers pointing upward. It is a gesture of veneration or respect to a greater being. Esoterically, it activates the heart and establishes more connection of the personality to the Soul.

Antahkarana: A cable of light that serves as a connector of consciousness from the incarnated person to the Higher Soul and vice-versa. The Higher *antahkarana* connects the Soul to the *Monad* or Spirit.

Astral: The substance of light that relates to feelings or emotions. It is a frequency between mental and etheric, made up of seven layers of vibration. Human beings have an astral body, as well as the Earth.

Atmic: A Sanskrit term pertaining to immortal consciousness. When applied as part of the human Soul, like "*Atmic* body", it is the aspect of will-power at the Soul level. If used as the "*Atmic* plane", it refers to the Earth's plane made up of the substance of higher will-power.

Aura: This is the general term for the invisible energy fields that contain and exude different types of consciousness and intelligences, generally constructed out of etheric, emotional, mental and spiritual vibrations

Bodhisattva: A Sanskrit term meaning a highly enlightened being who chooses to help and take upon himself/herself the karma of sufferings of other beings for the sake of redemption. Bodhisattva Maitreya refers to the Christ.

Buddha: A Sanskrit term meaning the awakened one or a being who has achieved enlightenment and has been freed from the cycle of human existence. If referring to the Lord Buddha, he is one of the Earth's few enlightened Buddhas, who incarnated as Gautama Sidhartha in Nepal in 563 BC. He was called Shakyamuni during the period of his life when he was a wandering ascetic.

Buddhic: A Sanskrit term for the consciousness of universal love, wisdom and intuition. When it is applied as part of a human Soul, like "*Buddhic* center", it is the aspect of love-wisdom and intuition at the Soul level. If used as the "*Buddhic* plane", it refers to the Earth's plane made up of the substance of universal love, wisdom and intuitional intelligence.

Causal Body: A part of the Soul that serves as the accumulator of the substance and essence of experience of an incarnated person's many lives. The byproduct of each incarnation created as substance of consciousness is deposited in the Causal Body. The contents of the Causal Body are diffused to the Higher Soul when one reaches the Sage level at which point the Causal Body is no longer needed.

Chakra: A Sanskrit term meaning "whirling wheel." The chakras are centers of energy that purify, revitalize and control the organs and systems of the physical body. It also controls the psychological, psychic and inner powers of the emotional and mental bodies of a human being.

Deva: A Sanskrit term meaning "shining one". They are generally referred to as Angelic Beings or invisible helpers who inhabit the inner realm. Their kingdom is in parallel evolution with humanity.

Egoic Lotus: An energy center shaped like a flower in this current cycle that connects the Soul to the incarnated personality. It is the converter and transmitter of the substance and essence of the experience of the incarnated person to be deposited in the Causal Body. It controls the flow of Divine life and light to the energy bodies of the personality.

Esoteric: Pertains to the hidden, deeper and universal aspects of truth and knowledge

Great Ones (Community of Saints, Society of Illumined Minds): Great enlightened beings who, in high places and development, formulate and send universal ideas and organize Divine strategies that are plans for the Earth and humanity. They are also called the Spiritual Hierarchy, which is composed of Saints, Sages and beings of very advanced development.

Ida: A Sanskrit term referring to the channels of the *kundalini* path regulating the passive or feminine aspect. It is affected by left nostril breathing.

Incarnated person or personality: The constitution of a human being that includes the physical body, etheric body, emotional body, mental body and the Egoic Lotus, an energy center that connects the Soul to its lower bodies.

Karma: A universal law. When interpreted at the human level, it is called the "Law of Cause and Effect" (i.e., what a person plants, so shall he/she reap).

***Kundalini* (Sacred Fire)**: The force inherent in nature and beings, including Earth and human beings, that awakens when stimulated by spiritual development or practices. It serves as a creative energy fertilizing and vivifying different chakras or energy bodies of the incarnated personality, the Soul and the Spirit. It is currently symbolized in its inner form by the caduceus of medicine and the many snakes at the back of the head in portraits of advaced yogis.

Macrocosm: The bigger whole where a minute fragment of existence belongs. The Earth is the macrocosm of the human being as the human being is the macrocosm of the cells in his or her body.

Meridians: The pathways of energy connecting the chakras and the many small centers of light in the energy bodies.

Microcosm: The fragment or miniature correspondence of a bigger whole. The human being is a microcosm of the Earth Being, and the Earth Being is a microcosm of the Cosmic Entity.

Monad (see also **Spirit**): The Divine Self or *Monad* is the one God within, but characterized with three divine qualities, which are a replica of the qualities of the planetary deity: Will and power, Love-wisdom and Active Intelligence.

Mudra: A Sanskrit word literally meaning "seal" or "sign". Esoterically, it is used as a special hand position and finger arrangement to seal or flow energies.

Nadis: A Sanskrit term that means "tube" or "vessel". They are the passageway of subtle energies flowing all around the aura of the human energy system.

Pingala: A Sanskrit word referring to one of the channels of the *kundalini* path regulating the masculine or active aspect. It is affected by right nostril breathing.

Prana: A Sanskrit term meaning "breath of life". Esoterically, it refers to the different levels of vitality circulating in the etheric chakras and aura. In its general usage, it pertains to the air vitality that comes with the oxygen in the air.

Shamballa: A Sanskrit word pertaining to a mythical power. Esoterically, it is the planetary center of power that serves as the head chakras of the planetary being and is the main source of Divine energy and power on Earth. It also refers to the group of the highest and most advanced beings on Earth.

Soul: It is an intelligent consciousness serving as a mediator between the Spirit and the incarnated personality. It serves as a transmitter and converter of Divine electricity and life from the Spirit to the incarnated personality.

Spirit (see also ***Monad***): The Divine Self or the God within, the "I am that I am" within every person, which allows him or her to become a true Spiritual human being

Spiritual Hierarchy: An integrated group of illumined beings, also referred to as the Great Ones, which creates the Divine Plan for the Earth and humanity, gives guidance and protection and dispenses power to all the kingdoms, especially humanity.

Sushumna: A Sanskrit term meaning "middle channel". It is the central energy meridian that serves as the major part of the *kundalini* to flow from its seat upward to the crown chakra.

Sutratma: A Sanskrit term meaning "thread-self". Esoterically, it is the life that descends from the Spirit to the Soul and then to incarnated personality. In this book, it is also known as the **Divine Electrical Cord**.

The Third Eye: The Third Eye is an advanced inner faculty of Saints and Sages that allows them to penetrate deeply into the realms of truth, with the additional faculties of discrimination, wisdom and intuition. It is an esoteric tool that is developed when the head chakras are very active and integrated, vivified by the *kundalini* energy and the Soul.

End Notes

[1] Leadbetter, Charles. *The Inner Life*. (1978: The Theosophical Publishing House, Wheaton, Illinois) p. 199 and Sui, Choa Kok. *Advanced Pranic Healing*. (1995: Samuel Weiser, Inc., York Beach, Maine) p. 26
[2] Huey, John. "Sam Walton" in *Time 100: American Legends* (New York: *Time* Books, 2001) p. 82
[3] Ibid, p. 80
[4] Tannen, Deborah. "Oprah Winfrey" in *Time 100: American Legends* (New York: *Time* Books, 2001) p. 151
[5] Ibid
[6] Gelernter, David. "Bill Gates" in *Time 100: American Legends* (New York: *Time* Books, 2001) p. 85
[7] Ibid, p. 86
[8] Lemonick, Michael D. "Edwin Hubble" in *Time 100: American Legends* (New York: *Time* Books, 2001) p. 100
[9] Ibid, p. 99
[10] Ibid
[11] Blake, William. "Auguries of Innocence" in *The Complete Poetry & Prose of William Blake.* Lines 1-4. (New York: Anchor Books, 1988) p. 490
[12] Kluger, Jeffrey. "Robert Goddard" in *Time 100: American Legends* (New York: *Time* Books, 2001) p. 97
[13] Schickel, Walter. "Walt Disney" in in *Time 100: American Legends* (New York: *Time* Books, 2001) p. 66
[14] Blake, William. "Annotations to the Works of Sir Joshua Reynolds" in *The Complete Poetry & Prose of William Blake.* Lines 1-4. (New York: Anchor Books, 1988) p. 647
[15] Matthiessen, Peter. "Rachel Carson" in *Time 100: American Legends* (New York: *Time* Books, 2001) pp. 105-107

Index of Sources of Selected Figures

1. Figure 9 (The Crown and Heart of Mary): www.ewtn.com/Devotionals/Heart/maryHeartsm.jpg
2. Figure 9 (Crown of the Christ): home.earthlink.net/~kylander/Christ_the_King.html
3. Figure 9 (Crown of Kings and Queens): Joe Hewitt, www.joehewitt.com/gallery/3dart/crown.htm
4 Figure 9 (Crown of Buddha): www.borobuddur.com/sbm/pix/bronze/buddha18.gif
5. Figure 10 (Heart and Alta Major Centers of Jesus and Mary): www.thesacredheart.com
6. Figure 10 (Pharaoh's Headdress): www.sis.gov.eg/tut/html/tut0302.htm
7 Figure 14: www.shalincraft_india.com/images2/thangka65.jpg
8. Figure 33 (7 Cobras): www.mandala.com.au/dham/chaitanya.html
9 Figure 33 (5 Cobras): www.mataji.org/jpg/vishnu.jpg
10. Figure 34 (Medical Caduceus): www.gcrio.org/edu/pcsd/caduceus.gif

Services and Works by Master Del Pe
and his international team

Organizations founded by Master Del Pe, with an international network of centers serving executives, professionals, groups and organizations through spirituality and performance coaching, training and seminars: GLOCEN® and ESOCEN®.

GLOCEN's strategies revolve around maximizing performance in five areas of life to transform success into fulfillment.

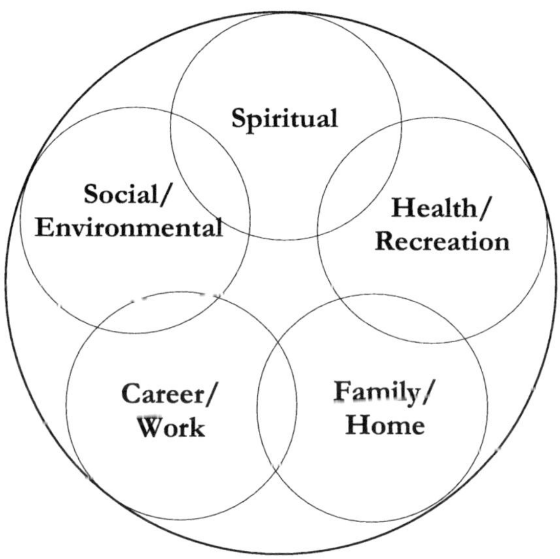

and

Maximizing balanced performance through five components:

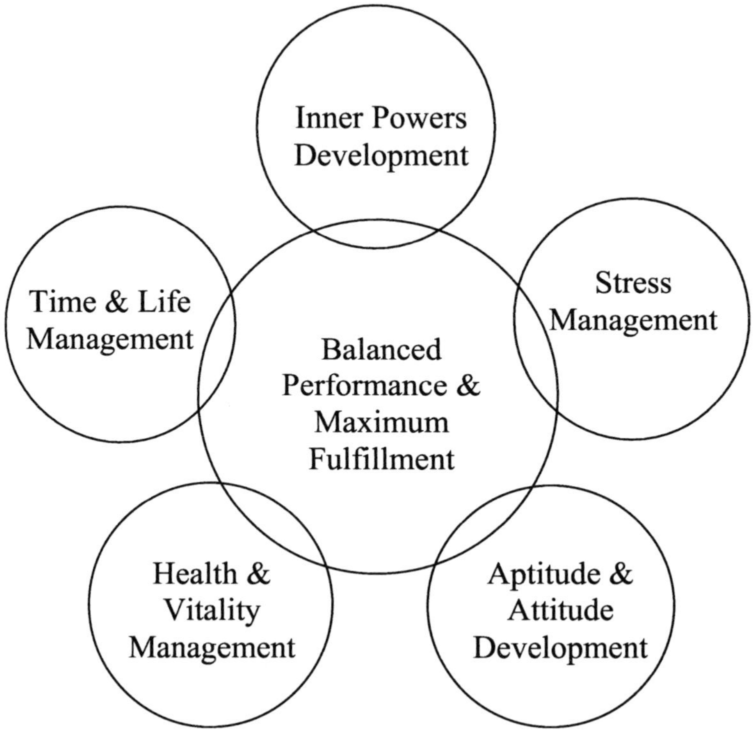

Master Del Pe has designed courses and training programs that implement these strategies:

Courses and Training Programs:

1. Inner Power Development to Maximize Performance™ (8 hours each/5 levels)
2. Managing Stress for Maximum Success™ (8 hours)
3. Integrated Time & Life Management™ (9 hours)
4. Mastering the Power of Decision-Making and Problem-Solving™ (8 hours)

5. Leadership in Action™ (How to bring balance in leadership) (8 hours each/2 levels)
6. Harmonizing the Light, Love and Power in Women™ (9 hours)
7. New Tools, Advanced Faculties and Inner Values for Executives and Professionals™ (2-day retreat or 16-hour seminar)
8. Inner Renewal and Stress Management Program™: On-going rejuvenation program and support for executives and professionals

ESOCEN specializes in training programs and seminars for individuals and their families, groups and organizations. ESOCEN focuses on spiritual transformation, continuous inner study, spiritual practices and group service as a way of life, applying the modernized teaching of the Ageless Wisdom.

Courses and Training Programs

1. Transformation of families and society:

- Quality Parenthood for Modern Families™ (8 hours)
- Quality Parenthood for Advanced Souls™ (8 hours)
- Stress Management for Families™ (8 hours)

2. Applied service and healing:

- Heal the Past, Transform the Present, Energize the Future™ (8 hours)
- Esoteric Healing and Metaphysics of Health™ (10 hours)
- Esoteric Healing Specialization I: MDP Smoking Cessation and Revitalization Program™ (10 hours)
- Esoteric Healing Specialization II: MDP Pain Control and Management Program™ (10 hours)
- Esoteric Healing Specialization III: MDP Stress Management and Healing the Side Effects of Stress Program™ (10 hours)
- Esoteric Healing Specialization IV: MDP Anti-Aging and Rejuvenation Program™ (10 hours)

- Advanced Esoteric Healing Course™ (10 hours)

3. Training on spirituality and inner transformation:

- Practical Spirituality from the Ageless Wisdom™ (8 hours)
- Life After Death and the Nature of the Inner World™ (10 hours)
- The New Psychology of Aquarian Discipleship™ (10 hours)
- The Science of Growth Towards the Light™ (10 hours)
- The 7 Types of Human Behavior and the 7 Forces of Creation™ (10 hours)
- Divine Alchemy and the Sacred Fires™ (10 hours each/3 levels)
- Mantric Yoga and the Science of Sacred Sounds™ (10 hours)
- Inner Powers to Maximize Performance™ (8 hours each/10 levels)

For Further Information or Services

GLOCEN and ESOCEN conduct coaching, consulting, seminars and training programs in in-house and public venues in over 15 countries. Inner Renewal and Stress Management Centers in several countries provide on-going support and nurturing for graduates and members. Our international team is dedicated to continue serving and assisting our clients on a personal and continuous basis after seminars or consulting engagements.

Please write or call:

GLOCEN: www.glocen.com • mdp@glocen.com
ESOCEN: www.esocen.com • mdp@esocen.com
President
P.O. Box 7947
The Woodlands, Texas USA 77387
Tel: (800) 352-6014 • (936) 273-9153
Mobile: (936) 520-2498
Fax: (936) 273-9230

Other Books and Products by Master Del Pe
Produced by MDP Global Resources®

Books (Year 2002 release):

1. *Inner Powers, Volume II: Manual for Spiritual Technology*

   A book for more advanced spiritual readers, requiring an open mind to dig into the deeper essence of the Ageless Wisdom teachings receptively. The reader is expected to have read *Inner Powers, Volume I* to receive maximum benefit from Volume II. It is a spiritual manual, bridging the traditional understanding of Ageless Wisdom topics into a modern understanding. The book offers:

   - A spiritual science that cultivates and cultures advanced spiritual practitioners to the next level, helping them achieve a higher level of conscious inner development
   - A practical method to construct the spiritual cable connection bridging the incarnated personality, the Soul and the Spirit, aiming at the greater integration among the three aspects
   - More detailed illustrations of the esoteric anatomy of the incarnated personality, the Soul and the Spirit, comparing the systems for early humans up to the Super Being level
   - A comparative study of the seven types of human behavior and how these different temperaments manifest as triggering factors in world affairs and the formation of cultures
   - A compilation of more than 4,000 years of world history elaborating on the shaping of spiritual paths and our modern life

- A technical update of the current configuration of the Spiritual Inner Government and how its departments influence world events
- A practical presentation of the deepest aspects of highest human nature, revealing a greater understanding of life's purpose, beauty and continuity

2. *Inner Powers, Volume III: Spiritual Strategies in the Period of Transition*

    - An advanced reference that demonstrates the work of the Inner Evolutionary Forces, creators of the past, present and future events of planetary life through the seven types of cosmic energies called the "seven rays", and their ramifications. This book presents an integration of both concrete and very abstract concepts on how humanity evolves through the different periods qualified by cosmic cyclic forces.
    - An esoteric study of the first 150-year cycle of the new Aquarian period serving as the transition into the golden age of humanity and marking the externalization of the members of the Spiritual Hierarchy
    - A comparative study of world trends and humanity's lifestyle qualified by the passing Piscean era, which will end in 2010, and the new Aquarian period, which is the incoming influencing power in the world
    - A technical illustration of the new configuration of the Aquarian models of the energy bodies, energy centers and *kundalini* of advanced humans and higher beings
    - The new Aquarian virtues, spiritual practices and group work for world servers and advanced groups of humanity during the period of transition

- The general schemes of the planetary esoteric forms shaping from years 2010-2160, the first cycle of the Aquarian period

CDs and Video

1. *MDP Inner Renewal™ Meditation CD* (English & Spanish): Item #1
    - A guided meditation to heal your past and improve your relationships
    - An effective method to correct, prevent and manage your stress weekly
    - A practical and simple technique to purify, revitalize and rejuvenate your body, emotions and mind
2. *Chantrams of Transformation™ CD*: Item #2
    - Seven tracks of esoterically sequenced sacred mantrams and sounds chanted to transform not only the listener but also the environment
    - The chanting of sacred mantrams enhances the virtues of love, compassion, inner peace, will-power and wholeness.
3. *MDP Namascar Meditation™ CD*: Item #3
    - An effective method to attain deep emotional calmness and peace of mind
    - A simple and easy-to-follow meditation facilitating the connection of the incarnated personality, the Soul and the Spirit
    - A fast way of developing universal oneness and spiritual balance
4. *MDP Body Therapy and Inner Power Development Program™ Video*: Item #4

- A synergy of simplified martial arts and yoga exercises and inner breathing techniques to heal the body, oxygenate the blood and detoxify and rejuvenate the whole body
- A fast method of increasing vitality and maintaining good stamina
- An effective technique of empowering and aligning your energy centers for improved instincts and emotional, mental and spiritual intelligence
- A short program that can help you maximize your performance in life

MDP Global Resources® is an organization that produces and publishes multi-media products and materials directed at improving performance and quality of life by applying the teachings of the Ageless Wisdom. It specializes in CDs, videos, books and pamphlets that offer practical methods and solutions to life's problems in order for individuals and groups to live and work harmoniously and to achieve inner balance, success and fulfillment.

For further information and to order products:

Vice President
P.O. Box 7947
The Woodlands, Texas USA 77387
Tel: (800) 352-6014 • (936) 273-9153
Mobile: (936) 520-2497
Fax: (936) 273-9230
Email: mdp@mdpglobal.com
Web: www.mdpglobal.com

MDP Global Resources Products:

- MDP Inner Renewal™ CD
- Chantrams of Transformation™ CD
- MDP Namascar Meditation™ CD
- MDP Body Therapy and Inner Power Development Program™ Video

Recommended Reading

The best results in our life come through balancing whatever we choose to do. This is also true in book study and learning. In our present age of information explosion, it is not easy to determine which books are more useful. Many available resources can be either too theoretical or excessively materialistic in approach. I recommend the following for your study and research on topics to bring balance, success and fulfillment to your life:

I. Personal and Business Growth and Success:

1. *The 100 Absolutely Unbreakable Laws of Business Success.* Brian Tracey (San Francisco, CA: Berrett-Koehler Publications, Inc, 2000)
2. *The 7 Habits of Highly Successful People.* Stephen Covey (New York, NY: Simon & Schuster: 1990)
3. Leadership books by John Maxwell (Nashville, TN: Thomas Nelson Publishers)
4. *Don't Sweat the Small Stuff* series. Richard Carlson (New York, NY: Hyperion)
5. *Executive EQ.* Robert K. Cooper and Ayman Sawaf (New York, NY: The Berkeley Publishing Group, 1998)
6. *If I Knew Then What I Know Now.* Richard Edler (New York, NY: The Berkeley Publishing Group, 1997)

II. Spirituality, Inner Sciences and Ageless Wisdom:

1. *Inner Powers, Volume II: Manual for Spiritual Technology.* Master Del Pe (The Woodlands, TX: MDP Global Resources, 2002)
2. *Inner Powers, Volume III: Spiritual Strategies in the Period of Transition.* Master Del Pe (The Woodlands, TX: MDP Global Resources, 2002)

3. Books by Master Choa Kok Sui (Pranic Healing and Inner Sciences), Institute for Inner Studies, P.O. Box 4903 Makati City Central Office, Makati City 1289, Philippines
4. *Autobiography of a Yogi.* Paramahamsa Yogananda (Los Angeles, CA: Self-Realization Fellowship, 1972) To contact the Self-Realization Fellowship: 3880 San Rafael Avenue, Los Angeles, CA USA 90065
5. *The Etheric Double* and *The Astral Body.* Arthur E. Powell (Wheaton, Illinois: Quest Books, 1987 and 1992)
6. *The Mental Body* and *The Causal Body and The Ego.* Arthur E. Powell (London, UK: The Theosophical Publishing House, 1967 and 1978)
7. *Spiritual Politics.* Corine McLaughlin and Gordon Davidson (New York, NY: Ballantine Books, 1994) To contact the Center for Visionary Leadership: 3408 Wisconsin Avenue NW, Suite 200, Washington, DC USA 20016
8. Esoteric books by Dr. Douglas Baker, Baker Publications, "Little Elephant", High Road, Essendon, Hertfordshire, England
9. Books by Torkom Saraydarian, T.S.G. Publishing Foundation, Inc., Visions for the 21st Century®, P.O. Box 7068, Cave Creek, AZ USA 85331-7068
10. Books by Michael D. Robins, University of the Seven Rays Publishing House, 128 Manhattan Avenue, Jersey City Heights, NJ USA 07307
11. *Living with the Himalayan Masters.* Swami Rama (Honesdale, PA: The Himalayan Institute, 1999) To contact The Himalayan Institute: RR1, Box 400, Honesdale, PA USA 18431-9706
12. Theosophical books by Helena P. Blavatsky, Annie Besant and C.W. Leadbetter. The Theosophical Publishing House, Adyar, India
13. Esoteric books by Alice A. Bailey. Lucis Publishing Company, 120 Wall Street, New York, NY USA 10005

Master Del Pe is an integration of the East and the West. His alchemy comes from his experience and training as an engineer, businessperson, master in martial arts and esoteric science, seminar leader, and spirituality and performance coach for leaders and groups. He has received extensive training with four advanced Sages and has experimented in depth with teachings of the Ageless Wisdom. He brings to his expertise a distillation of researched information and actual experiences living in many cultures and teaching in and visiting about 40 countries. He helped establish and manage 13 businesses and spiritual organizations in India, the Philippines, the United States, Central America, South America and Saudi Arabia. He is currently the founder and president of GLOCEN, ESOCEN and MDP Global Resources, United States-based organizations with international centers whose mission is to help professionals, executives, leaders and groups from all walks of life to maximize performance, success and fulfillment.